Techniques for Molecular Biology

Techniques for Molecular Biology

Editors

D. Tagu

INRA Rennes, UMR INRA/AGROCAMPUS BiO3P
BP 35327, 35653 Le Rheu cedex, France

C. Moussard

Service de Biochimie Médicale, Faculté de Médecine
Place Saint-Jacques, 25000 Besançon, France

SCIENCE PUBLISHERS
An Imprint of Edenbridge Ltd., British Isles.
Post Office Box 699
Enfield, New Hampshire 03748
United States of America

Website: *http://www.scipub.net*

sales@scipub.net (marketing department)
editor@scipub.net (editorial department)
info@scipub.net (for all other enquiries)

ISBN 1-57808-361-3

Library of Congress Cataloging-in-Publication Data

Principes des techniques de biologie moléculaire. English
Techniques for molecular biology/editors, D. Tagu and C. Moussard.
p.cm.
Translation of: Principes des techniques de biologie moléculaire. 2e ed. rev. et augm. Paris : INRA, 2003.
Includes bibliographical references and index.
ISBN 1-57808-361-3
1. Molecular biology--Technique. 2. Molecular genetics--Technique. 3. Plant molecular biology--Technique. I. Tagu, D. (Denis) 1961-II. Moussard, C. (Christian), 1952- III. Title.
QH506.P74413 2005
572.8--dc22

2005051569

Published by arrangement with INRA, Paris.

Translation of: *Principes des techniques de biologie moléculaire*
2èm édition revue et augmentée
French edition: © INRA, Paris, 2003.
ISBN 2-7380-1067-9

Published by Science Publishers, Enfield, NH, USA
An Imprint of Edenbridge Ltd.
Printed in India

CONTENTS

I. Definition 1

1. Structure and Expression of a Eukaryote Gene Coding for an mRNA and a Protein 3
2. Parameters of Description of a Genome 5
3. Sequencing of Whole Genomes 9

II. Vectors and Cloning 13

4. Restriction Enzymes 15
5. Electrophoresis of Nucleic Acids 19
6. Description of a Plasmid and a Phagemid 21
7. Description of a Bacteriophage and a Cosmid 23
8. Description of a YAC and Other Large Vectors 25
9. Molecular Cloning 27
10. Genetic Transformation of Bacteria and Yeasts 31

III. Labelling of Nucleic Acids and Hybridization 33

11. DNA Labelling 35
12. Molecular Hybridization 39
13. *In situ* Hybridization of mRNA 43

IV. DNA Libraries and Screening 47

14. Construction of a Genomic DNA Library 49
15. Construction of a cDNA Library 51
16. Screening of a Library 53
17. Differential Screening: Subtractive Libraries, AFLP-cDNA 57
18. Differential Display RT-PCR 61
19. Differential Screening by SSH: Suppression Subtractive Hybridization 65
20. Differential Screening by RDA: Representational Difference Analysis 69
21. EST: Expressed Sequence Tags 73
22. DNA Microarrays: DNA Microchips, cDNA Filters 77

V. Characterization of a Gene **89**
23. DNA Sequencing 91
24. PCR: Polymerase Chain Reaction 97
25. RACE: Rapid Amplification of cDNA Ends 99
26. Genome Walking by PCR 103
27. RT-PCR: Reverse Transcriptase PCR 105
28. *In vitro* Transcription 109
29. Determination of Transcription Initiation Site 111
30. Functional Analysis of Promoters 113
31. Gel Retardation 117
32. DNase I Footprinting 119

VI. Genetic Transformation of Eukaryotes **121**
33. Genetic Transformation of Plants by *Agrobacterium tumefaciens* 123
34. Direct Gene Transfer into Plant Protoplasts 129
35. Direct Gene Transfer by Biolistics 131
36. Genetic Transformation of Animal Cells 133
37. Cloning of Animals 137
38. Transient Expression 141

VII. Analysis of Gene Function **143**
39. Recombinant Proteins 145
40. Baculoviruses of Insects, Vectors of Expression of Foreign Genes 149
41. Yeast two Hybrid System 155
42. Site-Directed Mutagenesis 161
43. Mutation Complementation in Yeast 167
44. Knock-out of Genes in Yeast 171
45. Gene Tagging 175
46. RNA Interference 181

VIII. Polymorphism of a Genome **185**
47. Molecular Markers 187
48. Genetic and Physical Maps 193
49. PFGE: Pulse Field Gel Electrophoresis 197
50. RFLP: Restriction Fragment Length Polymorphism 201
51. RAPD: Random Amplified Polymorphic DNA 203
52. AFLP: Amplified Fragment Length Polymorphism 205
53. Retromarkers 207
54. SSCP: Single Strand Conformation Polymorphism 213

55. DGGE: Denaturing Gel Gradient Electrophoresis 215
56. SNP: Single Nucleotide Polymorphism 217
57. Simple Sequence Repeats (SSR): Microsatellites 219
List of Contributors 225

I

Definition

Profile 1

STRUCTURE AND EXPRESSION OF A EUKARYOTE GENE CODING FOR AN mRNA AND A PROTEIN

A eukaryote gene with mRNA has a coding sequence bordered by regulation sequences. The latter serve as signals to begin or end transcription of the gene by polymerase RNA II. Some of these sequences (e.g., the TATA box of promoter) are recognized by proteins called "general transcription factors" because they help this enzyme in the stages of initiation and elongation of transcription.

Other DNA sequences[1] are recognized by "specific transcription factors" that modulate the expression of genes in space (depending on the cell type), in time (during development), and/or under the effect of stress. The effect of these proteins on DNA triggers or inhibits the transcription of the gene.

The gene in the coding strand is made up of exon and intron sequences. These two types of sequences are transcribed (primary transcription), but the introns are eliminated during the splicing. The RNA is first modified in the 5′-P end (addition of a cap) and in the 3′-OH end (addition of adenines), then spliced (elimination of introns) before being transported into the cytoplasm. There, the ribosomes attach on the mRNA (messenger RNA) and, by the intermediary of tRNA (transfer RNA), the mRNA is translated into the corresponding polypeptide.

[1]They can be found downstream and/or upstream of the coding sequence.

Denis Tagu

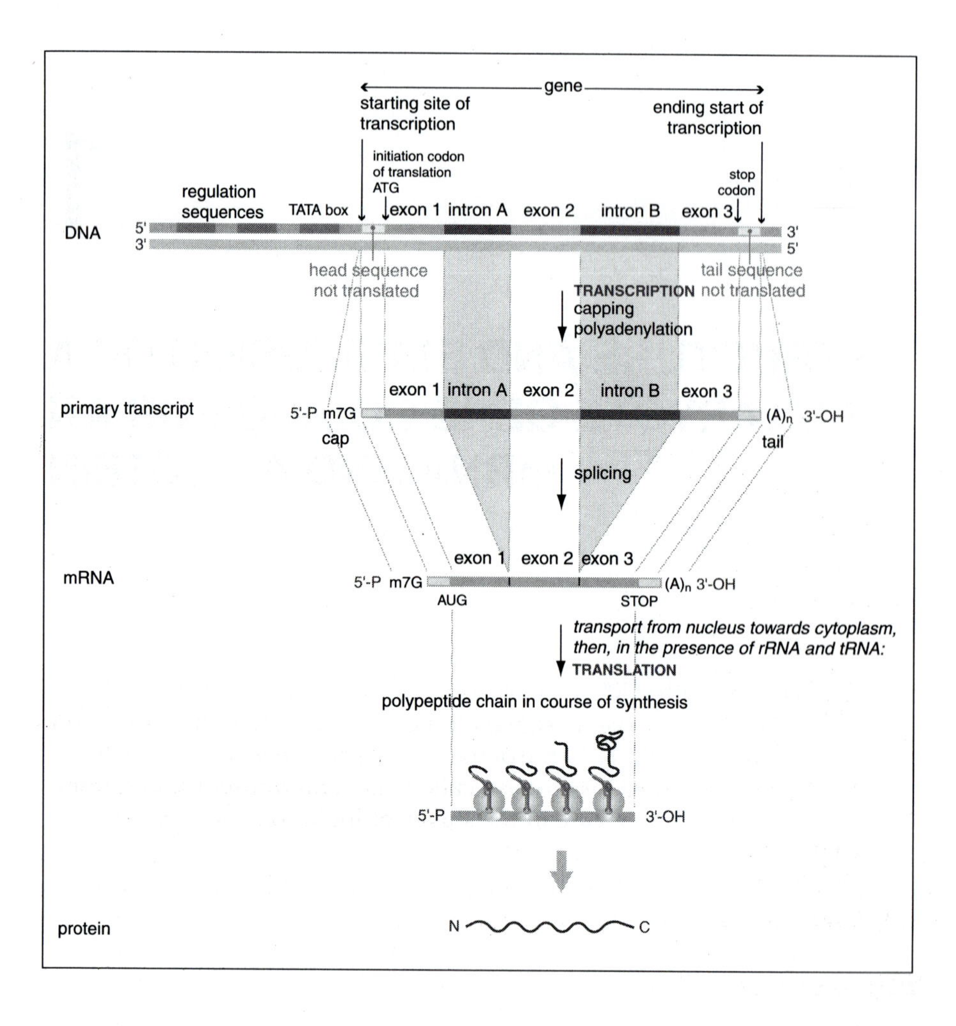
gene
starting site of transcription
ending start of transcription
initiation codon of translation ATG
stop codon
regulation sequences
TATA box
exon 1
intron A
exon 2
intron B
exon 3
DNA
5'
3'
3'
5'
head sequence not translated
tail sequence not translated
TRANSCRIPTION
capping
polyadenylation
primary transcript
5'-P m7G
cap
(A)n 3'-OH
tail
splicing
mRNA
exon 1
exon 2
exon 3
5'-P m7G
(A)n 3'-OH
AUG
STOP
transport from nucleus towards cytoplasm, then, in the presence of rRNA and tRNA:
TRANSLATION
polypeptide chain in course of synthesis
5'-P
3'-OH
protein
N
C

Profile 2

PARAMETERS OF DESCRIPTION OF A GENOME

Parameters[1]

- Quantity of DNA per 2C nucleus: about one picogram (1 pg = 10^{-12} g).
- Total length of DNA: expressed in base pairs (pb), kilobases (kbp), megabases (1 Mb = 10^6 bp).
- G + C percentage: percentage of cytosine and guanine bases in a known DNA sequence.
- Diameter of DNA: 20 Å. A DNA stretch of 1 Å length has a mass of $193/6 \cdot 10^{23}$ g.
- One base pair has a mass of 660/N, or $660/6 \cdot 10^{23}$ g.

Size of different genomes[1]

- Virus (DNA or RNA): from 10^3 to 10^5 bp, or 1 to 100 μm unrolled.
- Bacteria: 10^6 to 10^7 bp, or around 1 mm.
- Yeast: $1.2 \cdot 10^7$ bp.
- Higher plants:
  - Nucleus: 10^8 to 10^{10} bp, or around 1 m. The ploidy is generally 2n and the heredity is Mendelian.
  - Mitochondria: 1.8 to $6 \cdot 10^5$ bp. The length varies greatly. The ploidy is not determined and the heredity is cytoplasmic, maternal.
  - Chloroplasts: $1.5 \cdot 10^5$ bp, or 50 μm. The ploidy is about 10 × per chloroplast and there are 10 to 100 chloroplasts per cell. The heredity is cytoplasmic, maternal.

- Humans:
  - — Nucleus: $3 \cdot 10^9$ bp, or 1 m.
  - — Mitochondria: 16 to $20 \cdot 10^3$ bp.

The three types of sequences of the nuclear genome[1]

- "Single" sequences: a single or a few copies of a given gene per haploid genome (referred to as a small multigene family, e.g., genes coding for reserve proteins of Gramineae).These sequences contain genes of proteins.
- Moderately repeated sequences: 1000 to 100,000 copies per genome. They comprise genes coding for ribosomal RNAs (rRNAs), transfer RNAs (tRNAs), and possibly some protein genes.
- Highly repeated sequences: 100,000 to 1,000,000 copies. These sequences are not genes and may or may not be transcribed. Their role is not clear (e.g., satellite DNA).

Current data on genomes

Many prokaryote and eukaryote genomes have been sequenced or are being sequenced (see Profile 3). The following table gives some examples of physical parameters of genomes.

		Size (bp)	*No. of genes*
Eubacteria	*Agrobacterium tumefaciens*	$4.8 \cdot 10^6$ g	4,554
	Bacillus subtilis	$4.2 \cdot 10^6$ g	4,100
	Buchnera sp.	$0.6 \cdot 10^6$ g	564
	Enterococcus faecalis	$3.2 \cdot 10^6$ g	3,334
	Escherichia coli	$4.6 \cdot 10^6$ g	4,288
	Haemophilus influenzae	$1.8 \cdot 10^6$ g	1,709
	Helicobacter pylori	$1.6 \cdot 10^6$ g	1,566
	Lactococcus lactis	$2.3 \cdot 10^6$ g	2,566
	Listeria monocytogenes	$2.9 \cdot 10^6$ g	2,855
	Mycoplasma genitalium	$0.5 \cdot 10^6$ g	480
	Pseudomonas aeruginosa	$6.2 \cdot 10^6$ g	5,565
	Salmonella typhi	$4.8 \cdot 10^6$ g	4,600
	Sinorhizobium meliloti	$6.7 \cdot 10^6$ g	6,204
	Streptomyces coelicolor	$8.6 \cdot 10^6$ g	7,848
	Xylella fastidiosa	$2.6 \cdot 10^6$ g	2,766
	Yersinia pestis	$4.6 \cdot 10^6$ g	4,008
Archaebacteria	*Aeropyrum pernix*	$1.7 \cdot 10^6$ g	2,694
	Archaeoglobus fulgidus	$2.1 \cdot 10^6$ g	2,407
	Halobacterium sp.	$2.0 \cdot 10^6$ g	2,058
	Methanobacterium thermoautotrophicum	$1.7 \cdot 10^6$ g	1,869

(Contd.)

	Methanococcus jannaschii	$1.6 \cdot 10^6$ g	1,715
	Pyroaculum aerophilum	$2.2 \cdot 10^6$ g	2,605
	Pyrococcus abyssi	$1.7 \cdot 10^6$ g	1,765
	Sulfolobus solfataricus	$2.9 \cdot 10^6$ g	2,977
	Thermoplasma acidophilum	$1.5 \cdot 10^6$ g	1, 478

		Size	*No of chromosomes*	*No. of genes (est.)*
Eukaryotes	*Arabidopsis thaliana*	$125 \cdot 10^6$ g	5	25,500
	Caenorhabditis elegans	$97 \cdot 10^6$ g	6	19,000
	Drosophila melanogaster	$140 \cdot 10^6$ g	4	13,600
	Encephalitozoon cuniculi	$2.9 \cdot 10^6$ g	11	1,997
	Homo sapiens	$3100 \cdot 10^6$ g	21	35,000
	Oryza sativa	$420 \cdot 10^6$ g	12	40,000
	Saccharomyces cerevisiae	$12 \cdot 10^6$ g	16	6,100

Coding strand, sense strand: some definitions

DNA is made up of two anti-parallel and complementary strands. During the transcription of RNA, only one of the two strands is read and copied by the RNA polymerase. The RNA strand obtained is thus complementary to the template DNA strand that served as original. By definition, this template DNA strand is called antisense; the other strand, which has the same sequence as the mRNA, is the sense strand.

The mRNA is then read and decrypted into a protein; this is called translation (see Profile 1). By convention, the DNA sequence identical to the RNA (thus the sense strand) is called the coding strand; the writing of a gene on paper corresponds to the coding strand in its orientation from 5′-P towards the 3′-OH end.

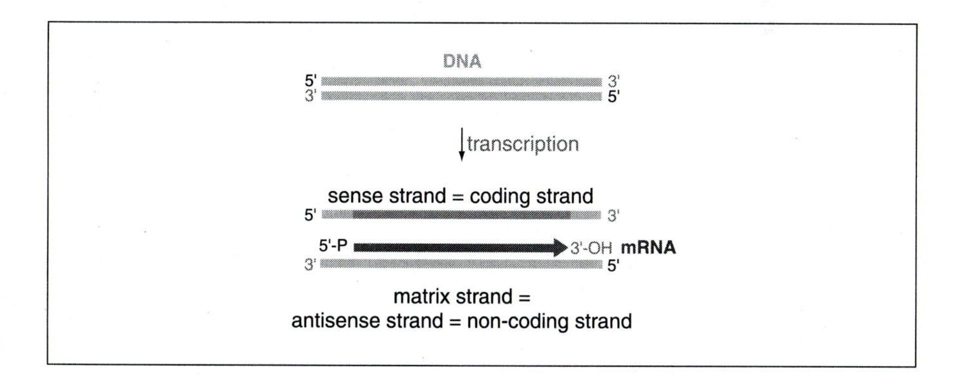

[1]According to Francis Quétier.

Profile **3**

SEQUENCING OF WHOLE GENOMES

The structure of a genome in its entirety can be understood through its sequencing. However, since genomes are several millions of bases (or megabases, see Profile 2) in size, it is necessary to combine the approaches of molecular biology, statistics, and informatics to be able to manage these millions of bases.

There are two methods of sequencing whole genomes. In both cases, the genomic DNA is first fragmented by either enzymatic methods (restriction enzymes) or by physical methods (ultrasound):

- The first sequencing method, referred to as hierarchical ordering, consists of classifying the genome fragments obtained before sequencing them.
- The second, called whole-genome shotgun, skips the preliminary trial phase and proposes to sequence each of the genome fragments in any order. By means of bioinformatics, the genome fragments can be reordered by overlapping their common sequences.

Hierarchical ordering

After extraction, the genomic DNA is fragmented, generally by processes of sonication, into manageable units (i.e., fragments that can be cloned in large vectors) of 50 to 200 kb, then cloned in an adapted vector, such as bacterial artificial chromosomes (BACs, see Profile 8). The number of clones must allow for a redundancy (representation) of 5 to 10 times the total genome. Clones are thus overlapped and ordered by hybridization of specific probes (molecular markers) or by analysis of restriction profiles, or more frequently by an ordering after sequencing and hybridization of BAC ends (500 to 600 bp). After the clones (thus genomic fragments) are ordered, the selected ones are individually fragmented, sequenced, and statistically assembled (alignment).

The advantages of hierarchical ordering are of two kinds: (1) It makes it easier to assemble sequences by means of data from physical and genetic maps and by means of overlapping of BACs. (2) It opens up possibilities for collaborative studies between several laboratories, each sequencing one region of the chromosome. The major disadvantage is the difficulty to assemble repeated sequences.

Whole-shotgun method

The whole-shotgun method is a "run-of-the-mill" method of sequencing the fragmented genome DNA, developed by the private company Celera Genomics on bacterial genomes, then on *Drosophila* genome and finally on human and mouse genomes. The steps are the following: Two to three libraries of random DNA fragments of different sizes are compiled. Numerous clones are sequenced and then assembled. The entire sequence is constructed through recovery and assembly.

The advantages of this method over hierarchical sequencing are speed and low cost. The disadvantage remains the problem of assembly, especially because of the significant presence of repeat sequences, particularly in mammal genomes.

Sequencing and assembly

For the two approaches described earlier, the fragments are sequenced and assembled in the same manner. The sequencing principle is based on the Sanger method (see Profile 23). The sequencing of a whole genome requires particularly effective DNA sequencers, as well as the assistance of robots that can complete thousands of pipetting and enzyme reactions in an automated manner.

After the recovery of raw sequencing data, the sequences must be assembled in order to reconstruct the initial sequence. Thus, all the sequences obtained must be ordered in relation to one another by the indication of overlapping ends, i.e., of terminal regions that have a chain of identical nucleotides (apart from sequencing errors). When two sequences overlap they are contiguous and form what is called a "contig", which designates two or several overlapping fragments. The aim of the project is to end up (by successive recoveries of all the contigs) with a single contig corresponding to the total sequence of the DNA molecule analysed (a chromosome). The error rate may reach 1%, keeping in mind that certain portions (such as repeat zone) are more difficult to sequence and assemble. The recovery of a sequence that covers 100% of a genome is particularly costly. For these reasons some genomes that are in the process of being sequenced are called "working drafts".

Annotation

Annotation is the final step: it is based on a statistical study of the sequences obtained and aligned. It consists of (1) locating the coding regions (genes); (2) determining the direction of transcription; (3) defining the coded proteins with indications of structure and function; and (4) identifying the gene families (paralogy[1]).

---

[1]The genes of a given genome that present a high rate of similarity and are probably derived from a single ancestral gene are called paralogues.

II

Vectors and Cloning

Profile 4

RESTRICTION ENZYMES

Restriction enzymes are of bacterial origin. They are peculiar in that they cleave double-stranded DNA molecules at specific sites in the sequence: these are endonucleases. Each restriction enzyme recognizes and cleaves a given nucleotide sequence. The name of the restriction enzyme comes from the name of the genus and species of the bacterium in which it was isolated.

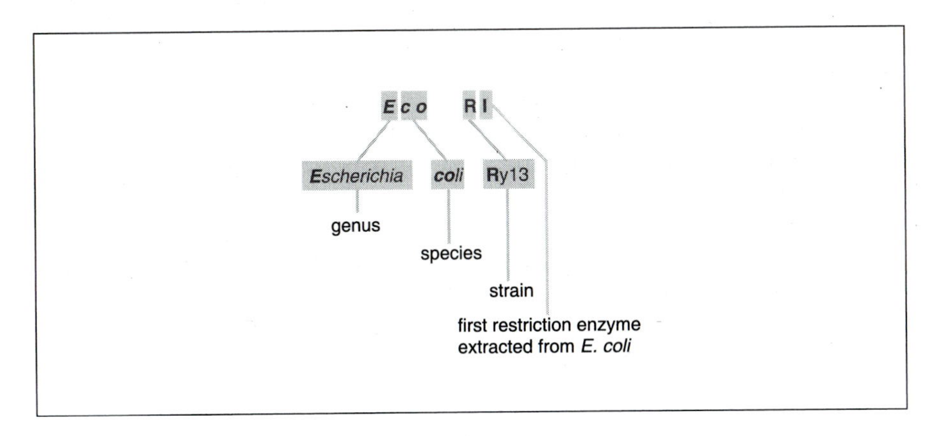

There are three types of restriction enzyme. Types I and III are complex proteins cleaving double-stranded DNA outside their recognition site. Type II restriction enzymes are indispensable tools in genetic engineering. They recognize a specific site of 4, 6, or 8 bp and cleave within this sequence, which is called the cleavage site. A particular characteristic of these sites is that they are generally palindromic, i.e., the same sequence is read on the two strands, but in reverse directions.

A large number of restriction enzymes are commercially available. They are generally supplied with their reaction buffer and the optimal temperature of activity is indicated. A single DNA fragment can be digested by two different enzymes. If they function in analogous conditions (concentration of salts, pH, temperature), the two digestions occur simultaneously. If not, the DNA fragment must be digested by the two enzymes successively, changing the experimental conditions between the enzymatic reactions.

The restriction enzymes used cleave within the sequence in two ways:

- Blunt cuts are obtained by cleavage at the same place in both the DNA strands. Example: the enzyme *Hae* III isolated from *Haemophilus aegypticus* cleaves double-stranded DNA at the GG/CC sequence.

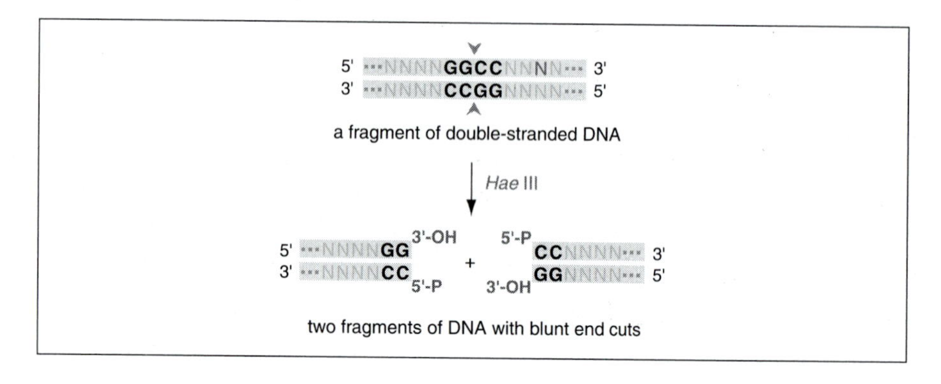

- Staggered cuts result when single-stranded ends are cleaved at complementary sequences called sticky ends. The following are examples:
  - — The restriction enzyme *Eco* RI isolated from the bacterium *Escherichia coli* cleaves double-stranded DNA in the palindromic sequence G/AATTC;

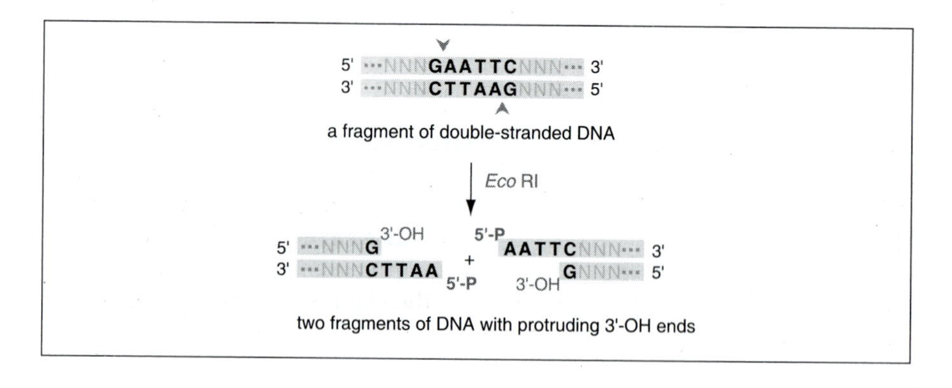

— The restriction enzyme *Pst* I isolated from the bacterium *Providencia stuarti* cleaves double-stranded DNA in the palindromic sequence CTGCA/G.

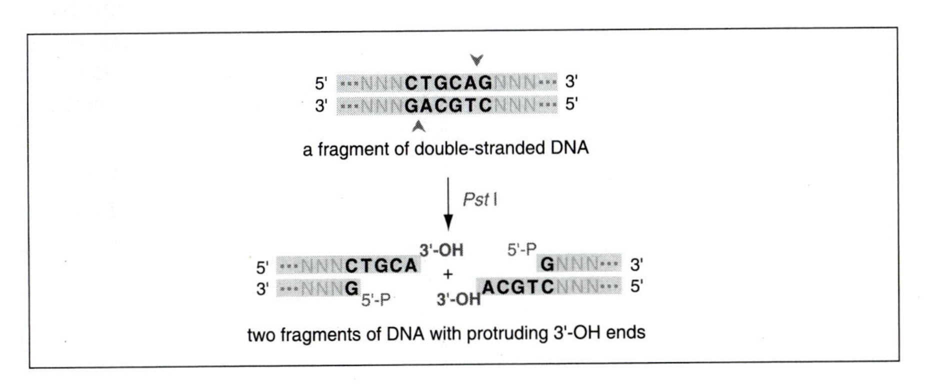

Profile **5**

ELECTROPHORESIS OF NUCLEIC ACIDS

The action of a restriction enzyme on a DNA molecule generates restriction fragments. These DNA fragments could be separated according to their size on an agarose or acrylamide gel. The mixture containing the DNA cleaved by the restriction enzyme is deposited at one end of the gel, which is then subjected to an electrical field.

The DNA molecules (negatively charged) migrate in the electrical field towards the anode. As they pass through the matrix of agarose or acrylamide, they are separated by size. The largest molecules are retained more easily than the smaller ones and migrate less quickly and thus less far in the gel. Acrylamide has a better ability to separate than agarose. However, the production of acrylamide gels is more difficult and the acrylamide is toxic.

In order to visualize the DNA fragments after electrophoresis, the gel is soaked in a solution containing ethidium bromide.[1] This molecule intercalates between the nucleic acid bases and emits a red-orange fluorescence when it is excited by ultraviolet light. The gel is then observed under a UV lamp and the DNA molecules complexed with ethidium bromide become visible. As the distance of migration is proportionate to the logarithm of the number of bases, it is possible to determine the size of restriction fragments obtained by comparing their electrophoretic mobility to that of DNA fragments of known size (size markers). There are also markers that can be used to estimate the quantity of double-stranded DNA in each fragment.

In some cases, the molecules to be separated have probably been marked by the incorporation of a radioactive isotope, which allows easy detection by

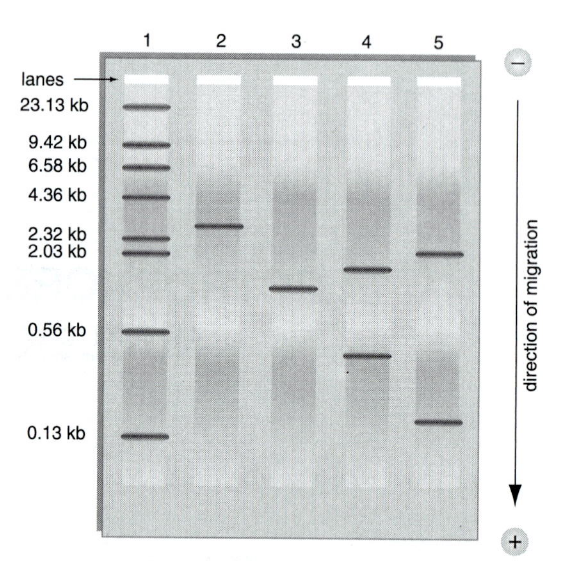

The samples are deposited in the lanes. After migration of DNA fragments, their size is determined by comparison of their distance of migration with that of fragments of known size (markers of molecular weight, lane 1). The size of markers of molecular weight indicated at left (DNA of phage λ digested by the restriction enzyme *Hind*III). kb, kilobase.

autoradiography: the energy particles emitted by the radio-isotope make an impression on a photographic film placed on the gel.[2]

[1]Ethidium bromide is a dangerous mutagen and must be handled in special conditions (with gloves and lab coat, and wastes recycled).

[2]Autoradiographic films are increasingly replaced by screens in which the signals are revealed digitally.

Profile **6**

DESCRIPTION OF A PLASMID AND A PHAGEMID

Plasmids and phagemids, which develop in bacteria, are widely used to manipulate gene fragments in any organism, by means of restriction enzymes (see Profile 4), PCR (see Profile 24), and molecular cloning (see Profile 9).

Plasmid (type pUC)

A plasmid is a small double-stranded DNA molecule (3 to 10 kb), circular, extrachromosomal, and capable of replicating (independently of the bacteria chromosome) in one bacterial cell and being transferred into another. Plasmids can be purified in large quantities because they multiply inside host bacteria.

The plasmids used in genetic engineering are natural plasmids that have been largely modified. Their genome comprises notably: (1) an origin of replication; (2) genetic markers or selection of transformed bacteria (gene for resistance to an antibiotic and gene coding for β-galactosidase); and (3) a polylinker. A polylinker is a sequence made up of a succession of sequences recognized and cleaved by different restriction enzymes. These sites are unique (they appear in only one place in the entire plasmid sequence) and are designed to receive the foreign DNA (see Profile 9).

Exogenous DNA molecules of 4 to 10 kb can easily be integrated in these plasmid vectors. On the other hand, larger molecules are difficult for plasmids to accept.

Phagemid (Bluescript type)

Phagemids are hybrid molecules between a plasmid and a phage. They are circular, double-stranded DNA molecules that can be obtained in single-stranded form under certain conditions. They have an origin of replication, at least one gene of resistance to an antibiotic, a polylinker, and a sequence from phage M13 containing the origin of replication that makes it possible to obtain the single-stranded molecule (by co-infection with a helper phage). In general, promoters recognized by RNA polymerases have been introduced upstream and/or downstream of a polylinker, in order to be able to produce RNAs by *in vitro* transcription (see Profile 28).

DNA fragments of around 4 kb can be introduced in these vectors, but it is sometimes possible to integrate inserts of around 10 kb.

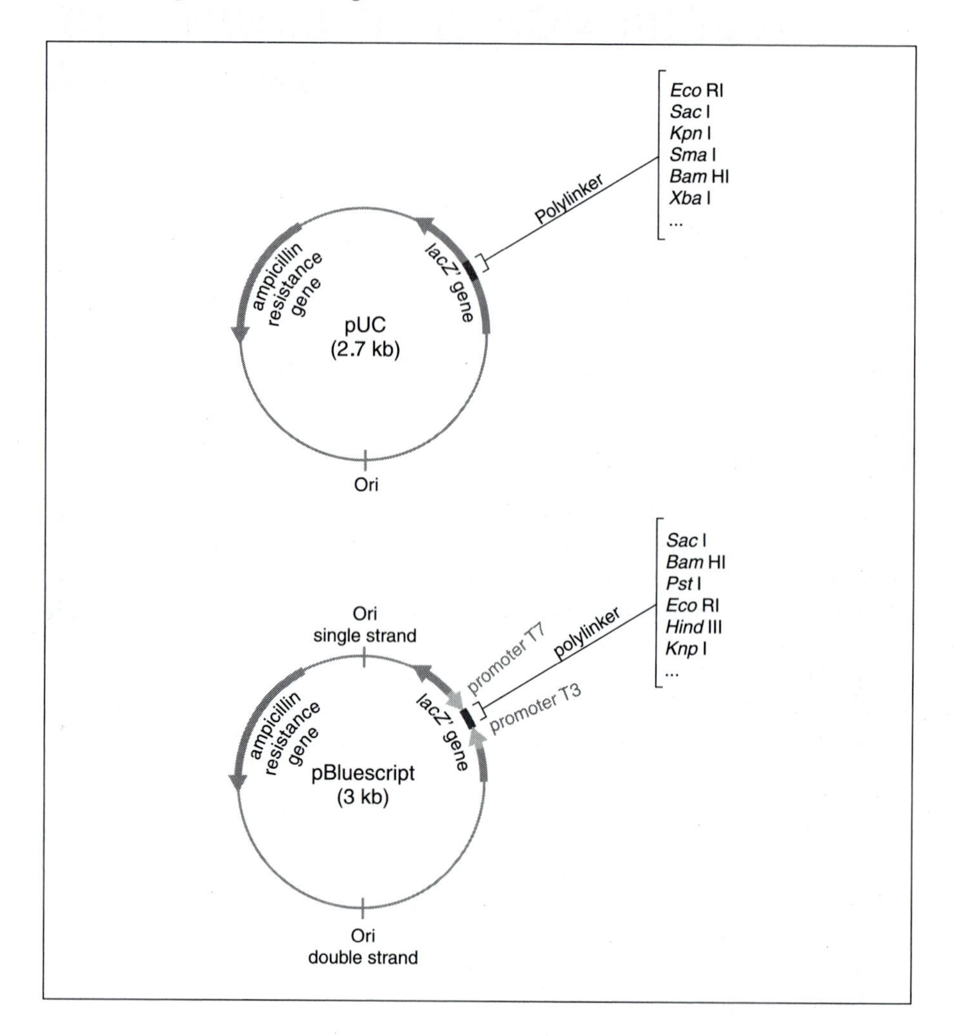

Profile 7

DESCRIPTION OF A BACTERIOPHAGE AND A COSMID

Bacteriophages and cosmids are used, as are plasmids, as cloning vectors, especially for construction of gene libraries (see Profiles 14 and 15). They can integrate larger DNA fragments than can plasmids.

Bacteriophage

A bacteriophage (or phage) is a bacterial virus. The most commonly used is bacteriophage λ. Its linear double-stranded DNA, around 50 kb, has sticky ends (*cos* sequence) that allow it to become circular inside the infected bacterium. The phage DNA is coated in a protein capsid shell. During the infection of a bacterium, the phage particle attaches to the outside of the bacterium and thus releases its DNA molecule inside the bacterial cell. The phage DNA will also replicate a very large number of times. The newly synthesized DNA molecules are encapsidated in new phage proteins within the bacterial cell. These are called virions. Then the bacterial cell lyses and releases thousands of phage particles identical to the phage that infected the bacterium. These new phages will in turn infect nearby bacteria.

The advantage of a phage vector over a plasmid vector is that its DNA accepts larger fragments of exogenous DNA (20 kb on average as opposed to 4–10 kb for a plasmid). However, because of the large size of the DNA, it is more difficult to manipulate *in vitro*.

In practice, phages are used as vectors for construction of cDNA or genome libraries (see Profiles 14 and 15). When the clones under study have been located, the phage DNA is purified and the extracted inserts are integrated into a plasmid vector that is easier to handle. This is called sub-cloning (see Profile 9). Sub-cloning can also be effected more simply by PCR (see Profile 24).

Cosmid

A cosmid is a plasmid (with its origin of replication, its marker genes, and its polylinker) in which are inserted the *cos* sequences necessary for encapsidation inside bacteriophage λ particles. The exogenous DNA fragments introduced in the cosmids must have a size between 35 and 45 kb. After ligation of a previously linearized cosmid and a foreign DNA fragment of suitable size (see Profile 9), the recombinant DNA can be encapsidated like that of a phage, then used to "infect" bacteria. Once inside the bacterium, the cosmid replicates like a plasmid, because it does not have the phage genes needed for the production of new phage particles. Cosmids are vectors used to construct genome DNA libraries (see Profile 14).

LAURENCE DAMIER AND DENIS TAGU

Profile **8**

DESCRIPTION OF A YAC AND OTHER LARGE VECTORS

Studies attempting to characterize the structure of whole genomes require the use of cloning vectors that allow the insertion of large DNA fragments. This makes it possible to handle a small number of recombinant vectors while covering the whole genome. YACs (yeast artificial chromosomes) and BACs (bacterial artificial chromosomes) are among the vectors that accept large fragments.

YACs are vectors constructed from chromosomal DNA sequences of yeast *Saccharomyces cerevisiae*. They have an origin of replication (ARS, autonomous replicating sequence), a centromeric site (CEN), and telomeric sequences (TEL) that allow them to behave like a chromosome once they are introduced in linearized form in the yeast. Several selection genes code for enzymes that allow selection of yeasts that have incorporated a viable vector (e.g., TRP1 and URA3[1]). A single cloning cite, located in the SUP4 gene,[2] allows the introduction of foreign DNA as well as the selection of recombinant vectors. In fact, the insertion of a DNA fragment interrupts the SUP4 gene, which allows the detection of strains that have incorporated a recombinant vector by appearance of a pink colour. These vectors are propagated in the yeast (see Profile 10).

Very large fragments of exogenous DNA (150 to 1000 kb) can be introduced into YACs. At present, the largest insert cloned is 2900 kb in size. These vectors are thus useful tools in the analysis of complex genomes but are subject to instability.

Other vectors also allow the cloning of large DNA fragments: e.g., bacterial artificial chromosomes (BACs) or P1-derived artificial chromosomes (PACs). These vectors, like YACs, can incorporate large DNA fragments. The BACs behave like very large plasmids, while PACs are handled like bacteriophages of type 1. BACs are increasingly used: in comparison to YACs, the cloning of DNA in BACs is easier and more

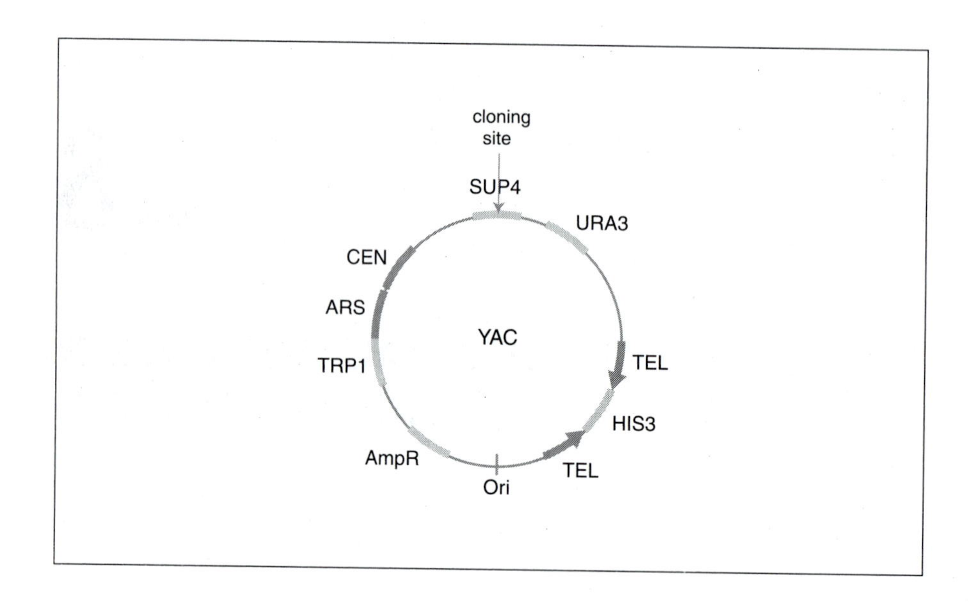

effective, and the clones are more stable in *E. coli* than in yeast. However, the size of fragments integrated does not exceed 300 kb.

Vectors	*Form of vector*	*Host*	*Size of insert (kb)*	*Major use*
Plasmids	Circular ds DNA	*E. coli*	0.1 to 10	cDNA libraries, sub-cloning
Bacterio-phages λ	Virus, linear ds DNA	*E. coli*	10 to 20	genomic and cDNA libraries
Cosmids	Circular ds DNA	*E. coli*	40	genomic libraries
Phagemids	Plasmid/phage hybrid	*E. coli*	4 to 10	cDNA libraries, sub-cloning
YAC	Artificial chromosome of yeast	Yeast	200 to 1000	genomic libraries
BAC	Artificial chromosome of bacterium	*E. coli*	100 to 500	genomic libraries
PAC	Virus, circular DNA	*E. coli*	80	genomic libraries

[1]The TRP1 gene coding for N-(5'-phosphoribosyl)-anthranilate isomerase is used as a selection marker: cells with this gene can grow in a medium without tryptophane. The URA3 gene coding for orotidin-5'-phosphate decarboxylase allows selection of cells that have this gene in a medium without uracil.

[2]The SUP4 gene codes for RNA-tyrosine.

Profile **9**

MOLECULAR CLONING

Cloning is based on the insertion of a fragment of exogenous DNA into a vector (plasmid, phagemid, bacteriophage, cosmid, YAC, BAC, PAC, etc.). The cloning vector is cleaved by a restriction enzyme that recognizes a single site (generally placed at the polylinker). Since this site is unique, the vector is thus linearized and has at each end one part of the DNA sequence recognized by the restriction enzyme.

The exogenous DNA of the donor organism is digested by the same restriction enzyme as that used for the linearization of the vector. The different fragments obtained thus also have at their ends part of the DNA sequence recognized by the restriction enzyme. When the open, linearized plasmid and the digested fragments of the donor DNA meet each other, there is hybridization (formation of hydrogen bonds) of complementary sticky ends, following the laws of complementarity of bases (A/T, C/G). An enzyme, ligase, allowing the formation of a covalent bond between these hybridized DNA fragments is then added to ligate the foreign DNA fragment to the plasmid. The effect of the ligase is to create phosphodiester bonds between the 5'-P and 3'-OH ends rendered adjacent by the hybridization of sticky ends of the fragment and the plasmid. The plasmid obtained, which is again circular, is called recombinant if it has integrated an insert.

When the plasmid DNA and the insert are digested by a single restriction enzyme, the insert could be integrated at random in either direction. In that case, the direction of insertion of the fragment must be determined by sequencing for example the recombinant plasmid (see Profile 23).

During the cloning that occurs after cleavage by a single restriction enzyme, two events of ligation may be produced: the ligation of a plasmid molecule with a molecule of the insert or the recircularization of the plasmid on itself, without the insert. Since the latter possibility is often more common, there must be an attempt to overcome it. In order to favour the insertion of a fragment of foreign DNA and thus enrich the population in recombinant plasmids, the recircularization of the vector on itself is prevented by treatment with alkaline phosphatase: the phosphate groups present at the 5'-P ends of the linearized vector are eliminated so thoroughly that the ligase can no longer catalyse the formation of the phosphodiester links between the two dephosphorylated ends of the plasmid. In contrast, it could catalyse the bond between a dephosphorylated end of the plasmid and a phosphorylated end of the insert.

Some less favourable cases may come up. When the plasmid and fragment of foreign DNA have been cleaved by a restriction enzyme generating blunt ends (Profile 4), the principle is the same, but the efficiency of the ligation is reduced. Nevertheless, the advantage of this approach is to permit the ligation of any ends, even if they do not have complementary sequences. This is the case especially for cloning into a plasmid of DNA fragments obtained by PCR[1] (Profile 24). In the case of cloning with "blunt

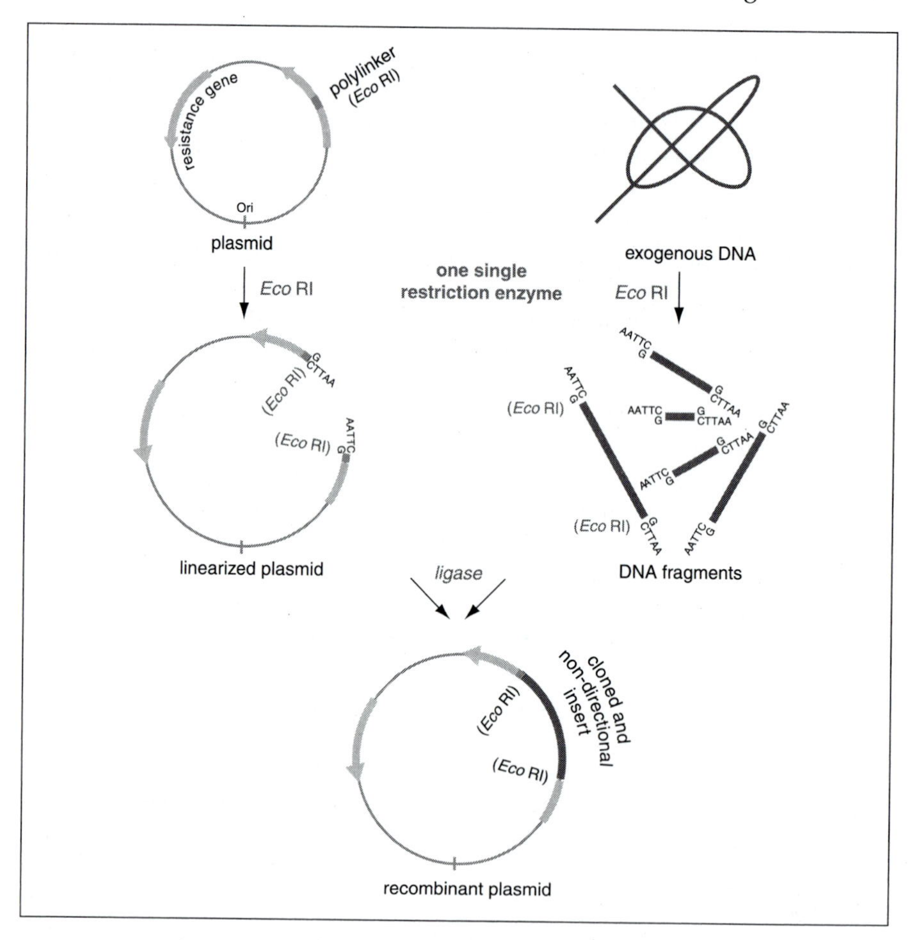

ends", a fragment of exogenous DNA could be inserted in either direction at random. The direction of insertion of a fragment must be determined by sequencing for example the recombinant plasmid (see Profile 23).

If the cloning vector and the exogenous DNA have not been digested by the same restriction enzyme and do not have blunt or compatible ends, it is essential to create blunt ends to insert the exogenous DNA into the vector. This is done, through the action of a DNA polymerase, by filling the protruding 5'-P ends by means of its 5' to 3' polymerase activity, or polishing the protruding 3'-OH ends by means of the 3' to 5' exonuclease activity of the enzyme.

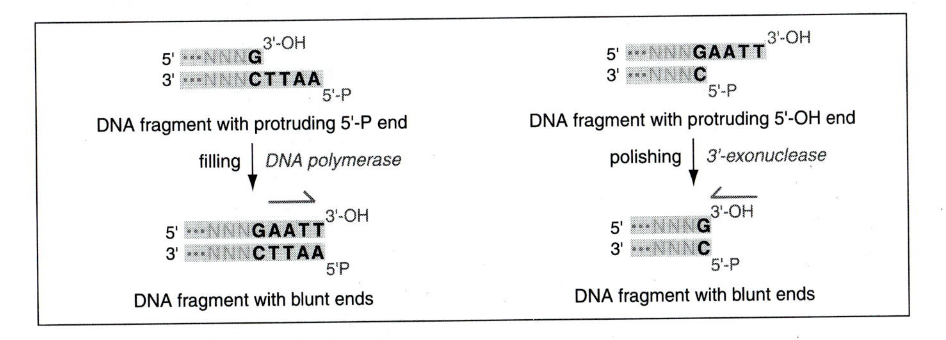

In some cases, the insert and the plasmid are digested by two different enzymes. The cloning is thus directional since only one direction of integration of the insert is possible.

---

[1]PCR generates double-stranded DNA fragments with a protruding A base in 5'-P. There are thus cloning vectors that are adapted, linearized, and in possession of a protruding T base (complementary to A) in 3'-OH. The clone yield is greatly increased.

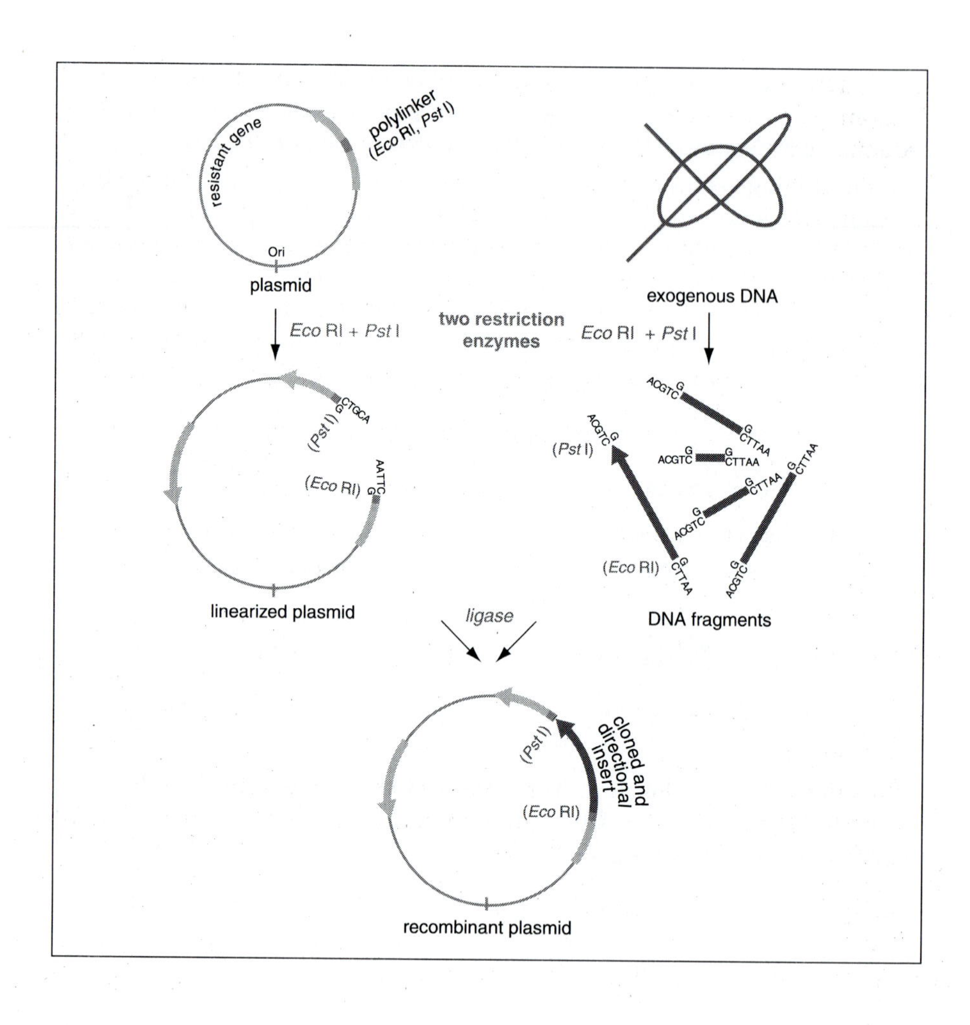

resistant gene
polylinker
(Eco RI, Pst I)
Ori
plasmid
exogenous DNA
Eco RI + Pst I
two restriction enzymes
Eco RI + Pst I
(Pst I)
(Eco RI)
linearized plasmid
(Pst I)
(Eco RI)
DNA fragments
ligase
(Pst I)
(Eco RI)
cloned and directional insert
recombinant plasmid

Profile 10

GENETIC TRANSFORMATION OF BACTERIA AND YEASTS

After DNA fragments are cloned in a vector, that recombinant vector must be incorporated into a microorganism in order to amplify, purify, or express the gene it contains. This incorporation is carried out by genetic transformation. The principle is to introduce into a bacterial cell (generally *Escherichia coli*) or eukaryotic cell (e.g., yeast) the recombinant vectors, i.e., those containing an exogenous DNA insert, obtained after cloning. When a recombinant vector is introduced into a new cell by genetic transformation, it can replicate there autonomously. In doing so, it replicates its DNA and also the DNA insert that it contains at the polycloning site.

Transformation of bacteria

In order to facilitate the penetration of DNA molecules through the cell wall and plasmalemma, bacteria in exponential growth phase are "weakened" (for example, by a calcium chloride treatment at 4°C). Bacteria that have been prepared in this manner, called "competent" bacteria, are mixed with a solution of plasmids to be incorporated. Then the plasmalemma of the bacterial cell is temporarily rendered permeable (transient formation of micropores) either by heat shock or by electric shock (electroporation[1]). The bacteria are then cultured on an agar medium containing the antibiotic corresponding to the resistance gene carried by the plasmid (see Profile 6). When this selective agent is applied, only the bacteria that have integrated a recombinant plasmid can grow. The recombinant plasmid will then multiply in the cell, at the same amplifying the cloned DNA sequence.

When a bacterium is transformed by a recombinant plasmid or phagemid, it grows on an agar medium and forms a colony. Each colony consists of a set of identical bacteria resulting from the division of a single bacterium. The maintenance of the plasmid in the recombinant bacteria is ensured by the selection pressure exerted by the presence of the antibiotic in the culture medium.

Transformation of yeasts

A hybrid DNA molecule can be introduced in the yeast *Saccharomyces cerevisiae* by two means. In the first method, spheroplasts (cells that no longer have cell walls) are prepared and incubated in the presence of recombinant plasmid DNA, polyethylene glycol (PEG), and calcium chloride. In the second method, intact cells with their cell walls are treated by an alkaline saline solution such as lithium acetate, which weakens the membranes, and are then incubated with DNA and PEG. The penetration of the DNA into the treated yeasts is stimulated by heat shock. The foreign DNA can be maintained in the transformed cell in a form integrated with chromosomal DNA by homologous recombination or in a replicative form independent of the chromosome. This capacity for independent replication is acquired from the sequences known as autonomous replicating sequences (ARS). The DNA molecules that do not have this ARS sequence must be integrated with a genome by homologous recombination in order to obtain yeasts that are transformed in a stable manner.

[1]Electroporation results in more efficient transformation.

III

Labelling of Nucleic Acids and Hybridization

Profile 11

DNA LABELLING

The hybridization of nucleic acids is a fundamental tool of molecular biology to detect a target DNA sequence. It is very frequently used to screen cDNA or genomic libraries, to study the organization of specific regions of the genome (by Southern blot), or to analyse the accumulation of transcripts in cells. The success of these techniques depends on the possibility of obtaining labelled DNA probes using radioactive or chemically modified nucleotides. Markers can be used to obtain probes labelled either uniformly (internal labelling) or at their ends (end labelling).

Internal labelling

Labelling by nick translation[1]

Internal labelling results from an *in vitro* polymerization of DNA in the presence of labelled nucleotides. This polymerization relies on the complementarity of two strands of DNA. The Klenow fragment of DNA polymerase I isolated from *E. coli* is capable, using the complementary strand as a template, of adding nucleotides to a 3'-OH end of DNA, which appears when one of the strands of a DNA molecule is cleaved. This enzyme also has a 5'-exonuclease activity that eliminates nucleotides on the 5'-P side of the cleavage. These two actions combined cause a movement of the cleavage site (cleavage translation) along the DNA. Before the action of DNA polymerase I, the cleavages are achieved by endonuclease. This endonuclease causes single-strand and random cleavages on the DNA.

If nucleotides are used that are labelled with ^{32}P with a high specific radioactivity, double-stranded DNA molecules are obtained in which some portions are strongly labelled (around 10^7 cpm/µg).

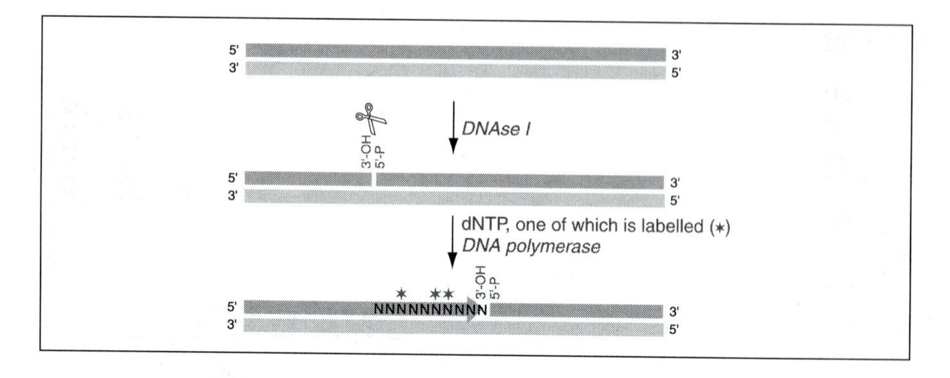

Labelling with random priming

A combination of hexanucleotides of random sequence is used as a source of primers for the *in vitro* synthesis of DNA from a double-stranded DNA molecule that has been denatured (the two strands separated by heat). If the combination of hexanucleotides is sufficiently heterogeneous, the probability of hybrid formation on any sequence of template DNA rendered single stranded is high. These double-stranded hybrid structures serve as a fixation point for a DNA polymerase (the Klenow fragment of DNA polymerase I of *E. coli* or increasingly often a *Taq* polymerase) that copies the complementary strand. The use of labelled nucleotides allows the synthesis of a radioactive molecule. The probes obtained have a very high specific radioactivity of about 10^8 to 10^{10} cpm/μg.

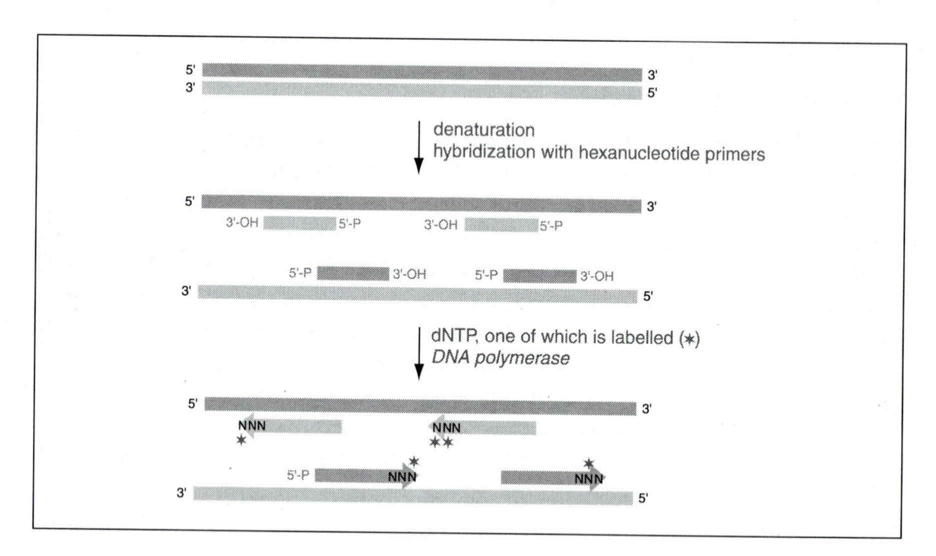

End labelling

End labelling is used for example for small molecules of some tens of nucleotides, such as an oligonucleotide. This technique is also useful when the objective is to get a polarity of labelling, as with DNase I footprinting (see Profile 32). The limit of this labelling is the introduction of a single labelled atom per DNA molecule, hence a rather low specific activity. Two different enzymes can be used to achieve the labelling of each end:

- Polynucleotide kinase of T4 bacteriophage catalyses the transfer of radioactive phosphate of [γ-^{32}P] ATP (ATP labelled with ^{32}P on the third phosphate) on the 5'-P end of a DNA fragment. This technique is especially applied to obtain labelled oligonucleotides that are used as probes for the screening of libraries (see Profile 14).
- The terminal transferase is also capable of adding a dideoxynucleotide (derivative of a nucleotide, see Profile 23) labelled at the 3'-OH end of a DNA fragment. The sequence of the labelled molecule is thus obtained and elongated from an analogue of the nucleotide to each of its ends.

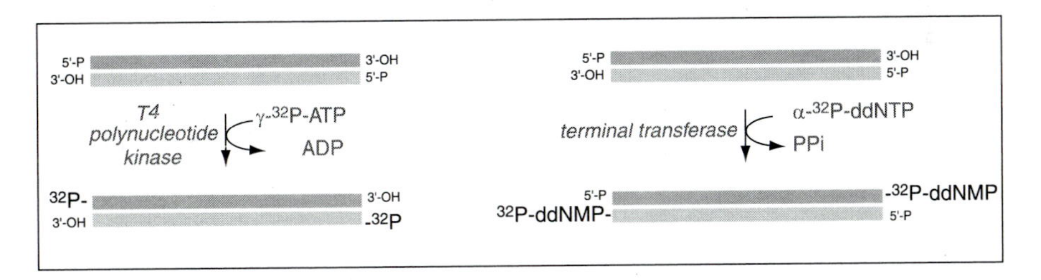

Non-radioactive labelling

Increasingly, radioactive nucleotides are being replaced by labelled nucleotides using non-radioactive molecules ("cold probes" as opposed to "hot probes"). There are various types of labelling techniques, with sometimes a higher detection sensitivity than that of radioactive probes, especially because of the use of fluorochromes. The molecules associated with the nucleotides include digoxygenin, biotin, and fluorescein or any other fluorochrome. The principles of labelling are the same as those described in the preceding paragraphs; only the detection techniques differ from one type of cold labelling to another. For example, digoxygenin is detected using specific antibodies and biotin using streptavidin.

---

[1]This type of labelling is increasingly replaced by random priming.

Profile **12**

MOLECULAR HYBRIDIZATION

Hybridization techniques are based on the physicochemical properties of nucleic acids: (1) the complementarity of the constituent bases of DNA (A/T, G/C) and RNA (A/U, G/C); (2) the reversibility of the process of separation of two strands of a DNA molecule (denaturation) and of re-association of the two strands (re-naturation). Denaturation is the dissociation of DNA strands (at 100°C or at alkaline pH) due to rupture of the hydrogen bonds that hold them together. The re-naturation process (re-formation of hydrogen bonds between the strands) is known as hybridization.

- If the two strands that hybridize present exactly the same sequences (100% similarity), the hybrid will be stable under conditions of significant stringency (see Profile 2).
- If the two strands do not have similar sequences, the hybridization will not occur.
- When there is partial similarity of the two strands, the hybrid will be unstable under conditions of high stringency but stable under conditions of low stringency. An example of a condition of high stringency is a medium similar to water (very low salt concentration) and high temperature (> 5°C). The temperature at which the two DNA strands separate is specific to each sequence (*Tm*). If the molecule studied is rich in G and C bases, the double-stranded molecule will be stable, because the G/C pair has three hydrogen bonds, while the A/T pair has only two.

Molecular hybridization can be used to locate or study a specific DNA or RNA molecule in a heterogeneous population. The principle is to use a radioactive or chemical tracer (see Profile 11) to label the known and purified DNA sequence that needs to be located in a population. This labelled single-stranded fragment constitutes the probe. The target molecules rendered single stranded (heterogeneous population of denatured nucleic acids) are fixed on a hybridization membrane made of nylon or nitrocellulose.[1] The single-stranded probe and the denatured target, which is also single stranded, are put into contact, the target on one membrane being incubated in a solution containing the probe, under conditions that allow hybridization. When the probe molecule recognizes its homologue in the population of target molecules, hybridization occurs and the hybrid becomes labelled by the presence of the probe. The membranes are rinsed suitably to eliminate any non-specific hybridization and retain only the desired probe/target hybrids. In the case of a radioactive label, the hybrids are located by placing the membrane in contact with an autoradiographic film that is then developed just like a photographic film.[2] In the case of chemical labelling, the molecular hybrid is often revealed by a colorimetric test on the membrane (see Profile 11).

Various hybridization techniques

DNA/DNA hybridization or Southern blot

The probe is made up of a DNA molecule. The target is also a DNA molecule such as a nuclear DNA, plasmid DNA, or phage DNA, a cDNA library, or fragments of target DNA sorted by size by means of agarose gel electrophoresis.

These target DNA molecules are denatured, then transferred by capillary action on to a nitrocellulose or nylon membrane. After the membrane comes into contact with the labelled probe under conditions favourable to hybridization, rinses are carried out in order to reveal after autoradiography labelled bands corresponding to DNA fragments complementary to the probe. One application of the Southern blot test is the diagnosis of genetic disease in the context of restriction fragment length polymorphism studies (see Profile 50).

RNA/DNA hybridization or northern blot

The probe is made up of DNA, but the target is an RNA. RNA molecules are isolated from a tissue and purified, then separated by electrophoresis on an agarose gel. The RNA molecules complementary to the probe are identified after transfer to a nylon or nitrocellulose membrane that is then subjected to hybridization. This type of DNA/RNA hybridization makes it possible to evaluate the size of the mRNA studied as well as the rate of accumulation of an mRNA population in a given tissue (leaf, root, stem) at a given point in its development (different phases of embryogenesis) or under the effect of a biotic or abiotic stress.

Dot-blot and slot-blot

The probe is made up of DNA, while the target may be single-stranded DNA or RNA. Unlike techniques of the Southern or northern type, the target

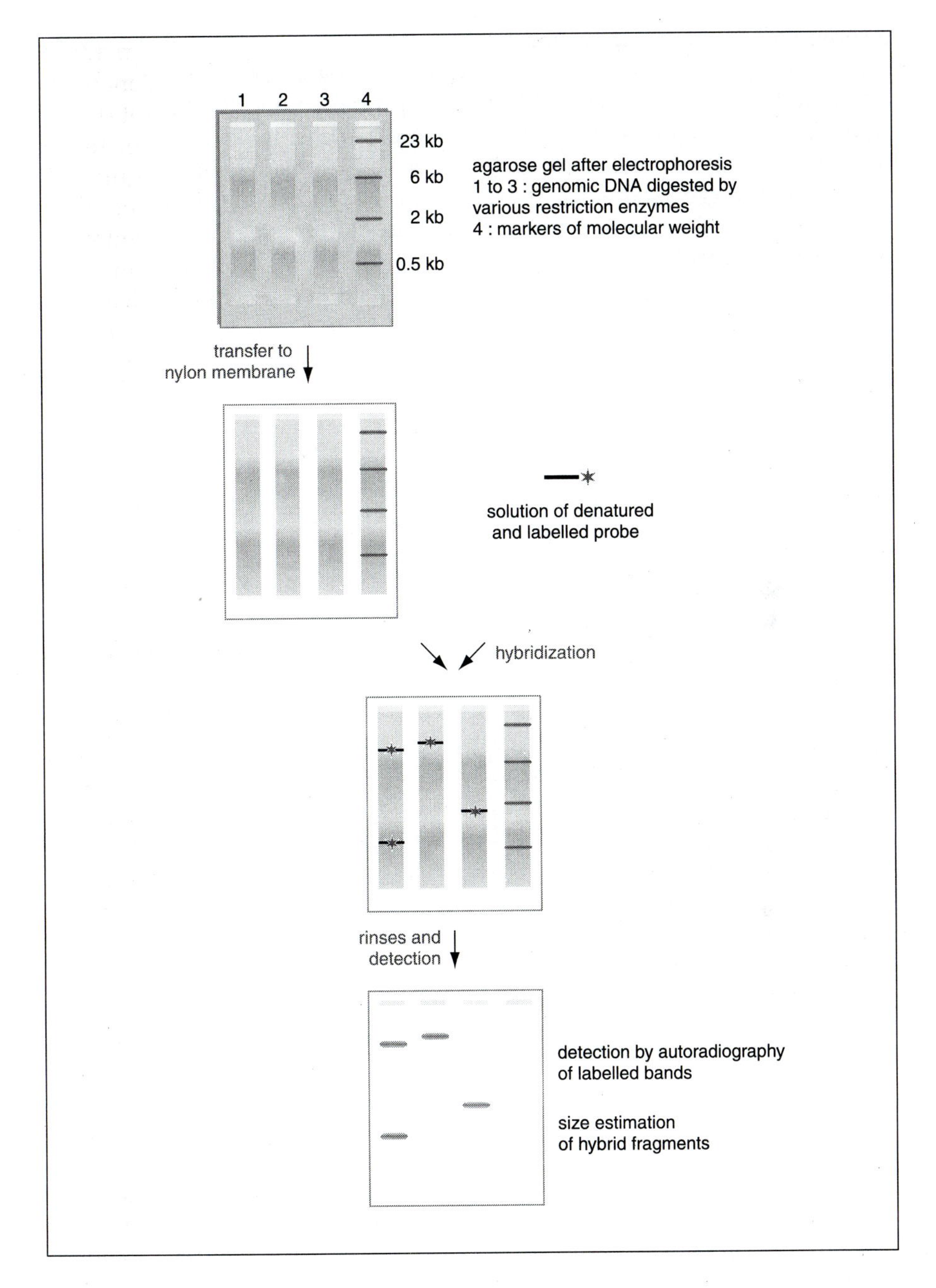

nucleic acids are not separated by electrophoresis. They are directly deposited on a hybridization membrane at a known concentration either in the form of a dot or in the form of a slot. Equipment adapted to these techniques is available on the market. After hybridization between the probe and the target, the intensity of the radioactive patch obtained reflects the concentration of the nucleic acid studied (probe) among the target nucleic acids. This technique is faster to use than experiments of the Southern or northern type. However, some of the probe DNA must be highly specific to the desired target sequence, because the absence of sorting of target molecules according to size (by electrophoresis) prevents the detection of non-specific hybridization.

These approaches are actively used in the field of transcriptomics, cDNA filters, and DNA microchips (see Profile 22).

---

[1]Nitrocellulose membranes allow less error than nylon membranes, but they are more fragile.

[2]Autoradiograph films are nowadays often replaced by autoradiographic screens in which the signals are revealed digitally.

Profile 13

IN SITU HYBRIDIZATION OF mRNA

A tissue is composed of several types of cells, each expressing a set of genes. Thus, two adjacent cells in a single tissue may not express the same gene. Sometimes the location of the gene transcript must be known for its mode of expression to be understood.

The principle of *in situ* hybridization of mRNA consists in identifying on a section the cells of a tissue accumulating a given population of mRNA. This technique is based on molecular hybridization. The target molecules are mRNAs present in the cells. The tissue or organ is first fixed chemically, by a process using paraffin, sliced into fine sections (around 5 to 10 μm thick), then put into contact with the probe. The probes used are usually labelled RNA (riboprobes). The riboprobes are created by *in vitro* transcription (see Profile 28) of a strand complementary to the target mRNA in the presence of a labelled UTP. This UTP is labelled either radioactively (generally with ^{35}S UTP and more recently with ^{33}P) or chemically (digoxygenin).

In the case of a radioactive probe, a liquid and translucent photographic emulsion is then run over the section after hybridization and rinse. This translucent emulsion will solidify, following all the contours of the section. The radioactivity can locally print the film and the section is examined under an optical microscope. There is superimposition of cellular structures and precipitation of silver grains.

The chemical labelling of ribosomes usually involves UTP coupled with digoxygenin (Dig), a steroid haptene. After hybridization and rinse, the RNA/RNA hybrid is detected using an anti-digoxygenin antibody coupled with alkaline phosphatase (ALP). The enzymatic activity carried by the antibody is detected by addition of a substrate forming a coloured precipitate. When the slides are observed under microscope, this precipitate can be seen at the cellular level and the location of the transcript deduced from it.

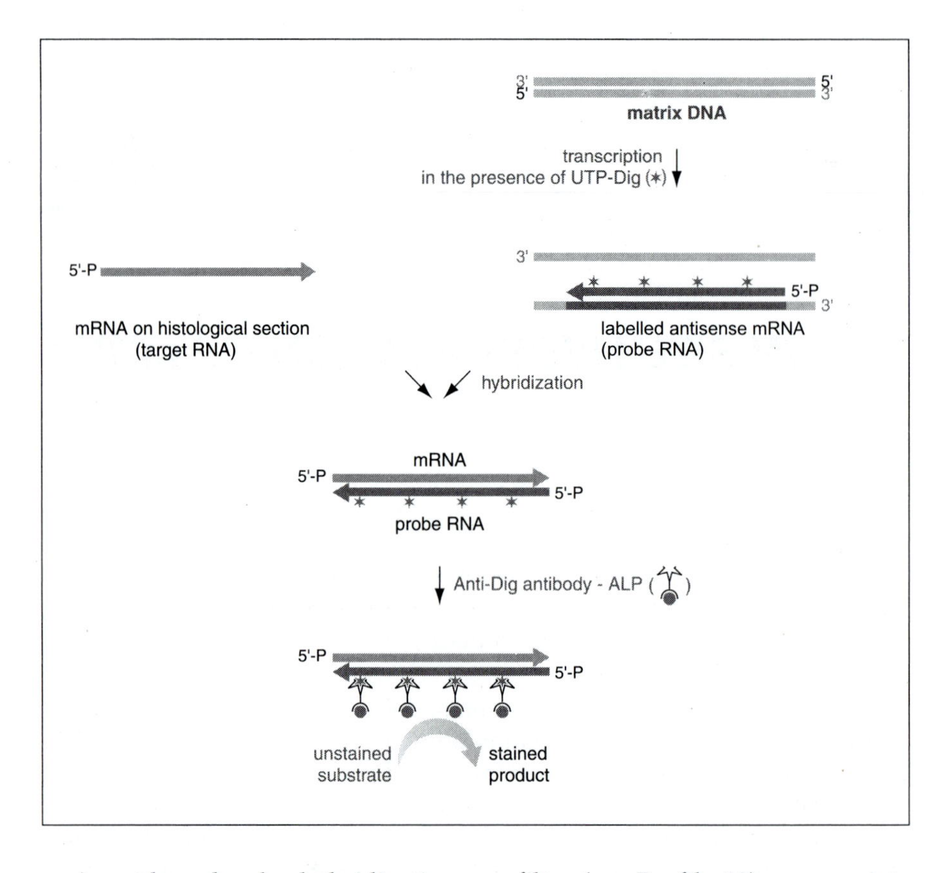

As with molecular hybridizations on filter (see Profile 12), appropriate rinses can be used to eliminate any non-specific hybridization. The hybrids are then revealed and the sections are examined under optical microscope. They can be stained to delimit the cellular structures and the cells accumulating the desired mRNA can be detected. The illustration below is of a plant root section.

In situ hybridization of mRNA has proved effective in studies of genes regulated during embryogenesis of *Drosophila* and has allowed the delimitation of sectors in which differential accumulation of mRNA (e.g., *nanos* and *bicoid*) generates differentiation of various tissues in the embryo. In the case of very young tissue, translucent and made of a small quantity of different tissues, *in situ* hybridizations can be carried out directly on the whole tissue without the use of histological sections.

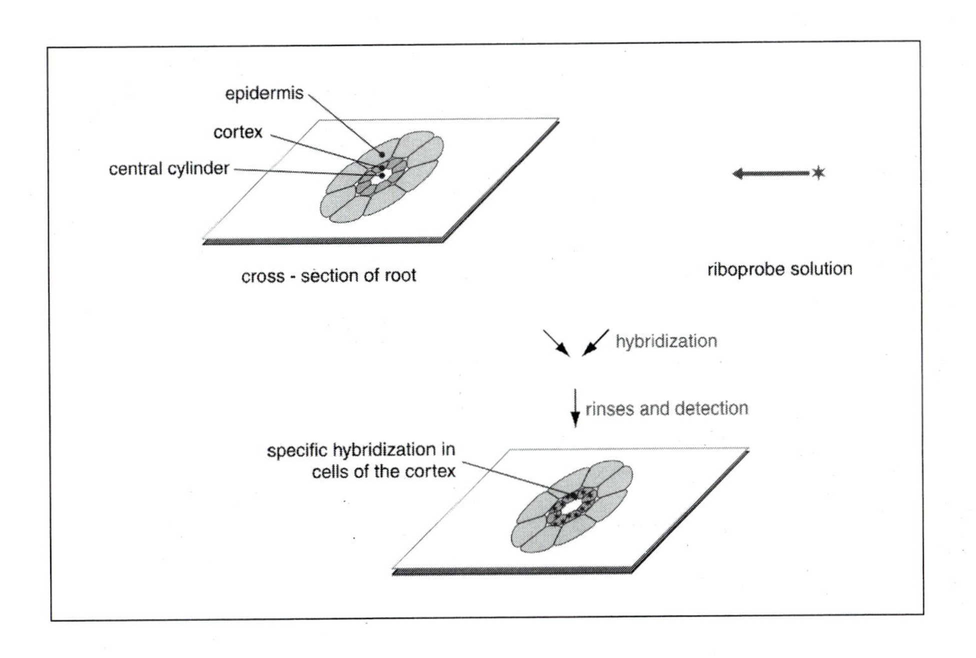
epidermis
cortex
central cylinder
cross - section of root
riboprobe solution
hybridization
rinses and detection
specific hybridization in
cells of the cortex

IV

DNA Libraries and Screening

Profile 14

CONSTRUCTION OF A GENOMIC DNA LIBRARY

Obtaining a genomic library is necessary, for example, when a particular genomic sequence is to be characterized, the physical map of a genome is to be established, or part or all of a genome is to be sequenced on a large scale.

A library of genomic DNA is a set of recombinant vectors each containing a different part of the genome of the species being studied. The principle consists of the following: (1) isolating the genomic DNA from the species being studied; (2) digesting that DNA using a restriction enzyme that allows the liberation of restriction fragments of a length compatible with the chosen vector;[1] (3) cloning the fragments in a cloning vector; (4) integrating the recombinant vectors in a microorganism in order to multiply them. The phage vectors (see Profile 7) are better adapted to cloning of a given genomic sequence, while the cosmids (see Profile 7) and YACs or BACs (see Profile 8) are chosen to characterize a genome. The library obtained is thus made up of millions of recombinant clones.

Since genomic DNA is considered identical in all the cells of an individual, the tissue from which it is taken is not important. For example, a genomic library constituted from DNA from the roots of a particular individual is theoretically identical to one constituted from DNA taken from the leaves of that same plant.

[1]Very small fragments of DNA must sometimes be eliminated so as not to saturate the library with small clones. This is done by incomplete digestion of the genomic DNA in order to obtain fragments of sufficient length.

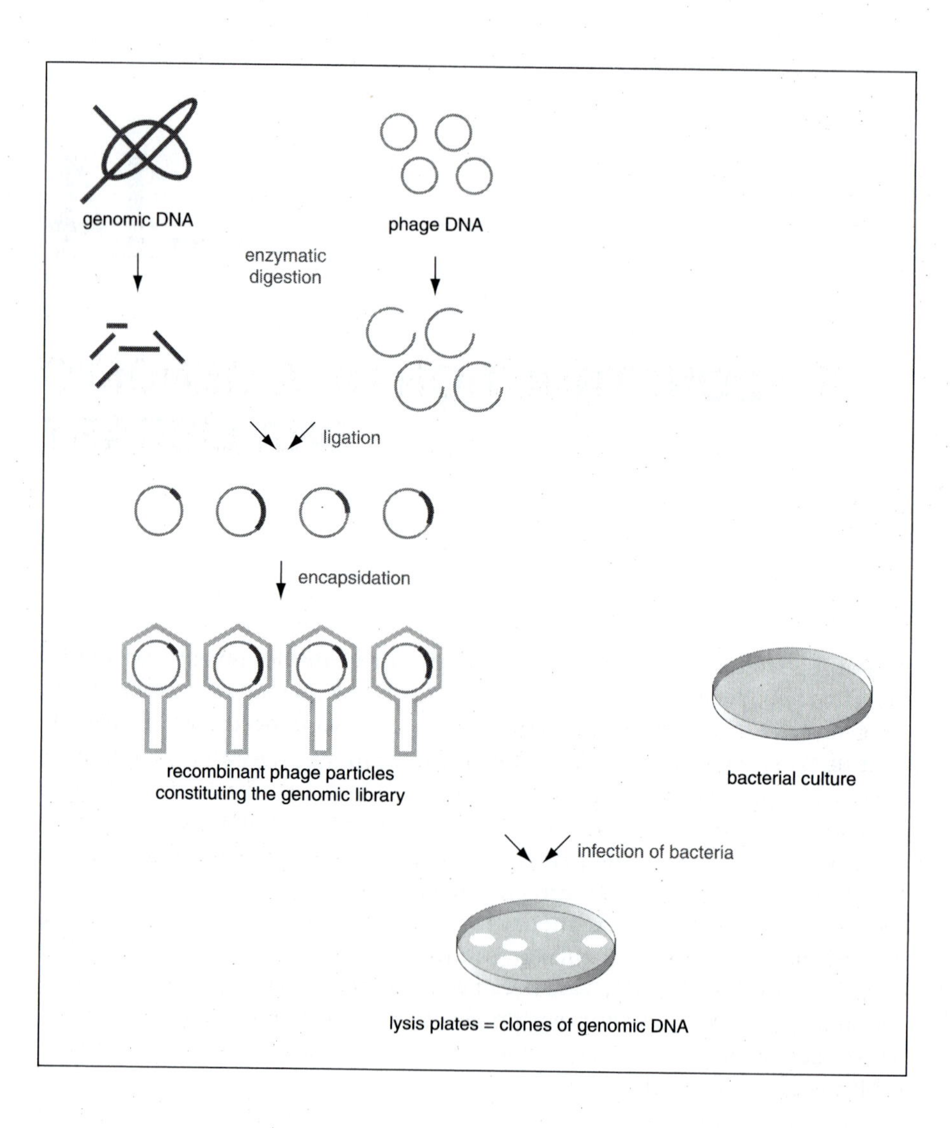
genomic DNA
phage DNA
enzymatic digestion
ligation
encapsidation
recombinant phage particles constituting the genomic library
bacterial culture
infection of bacteria
lysis plates = clones of genomic DNA

Profile 15

CONSTRUCTION OF A cDNA LIBRARY

Populations of mRNA accumulated in a given tissue are representative of that tissue. It is thus very important to be able to characterize the genes expressed in a given organ. However, the manipulation of mRNAs is a delicate process (small quantity, sensitivity to nucleases) and they must be transformed into double-stranded DNA that can be easily manipulated and adapted to cloning in bacterial vectors. A library of cDNA (complementary DNA) is thus representative of populations of mRNA present in a given tissue at a particular stage of its development. Unlike a genomic library, a cDNA library is specific to a tissue. It can be considered a "snapshot" of populations of mRNA represented in an organ.

The construction of a cDNA library is carried out in several steps. The principle is based on (1) extraction of RNA and sometimes the purification of polyadenylate mRNA of the organ (for example, by chromatography of affinity on a polyT column)[1]; (2) copying of these mRNAs into complementary single-stranded DNA by the action of reverse transcriptase; (3) specific elimination of mRNA by RNase H or NaOH; (4) synthesis of the second strand of DNA by a DNA polymerase; (5) ligation of oligonucleotides (adaptors) to create restriction sites; (6) ligation in a cloning vector (plasmid or phage); (7) integration of recombinant vectors in a bacterium. The cloning vectors used for the construction of cDNA libraries are either plasmids or phages (see Profiles 6 and 7). The synthesis of the second strand of DNA and ligation of adaptors are now frequently carried out using PCR[1] (see Profile 24).

The synthesis of a strand of cDNA is primed generally by fixation of a short polyT sequence on the poly(A) end of the mRNA. Another approach is to use as primer for the synthesis of a strand of cDNA by reverse transcriptase a combination of synthetic hexanucleotides corresponding to many different sequences and hybridizing at random on the mRNA (see Profiles 17 and 51). However, since by definition the priming of the synthesis of cDNA is done on the inner parts of the mRNA, the cDNAs correspond only to fragments of mRNA.

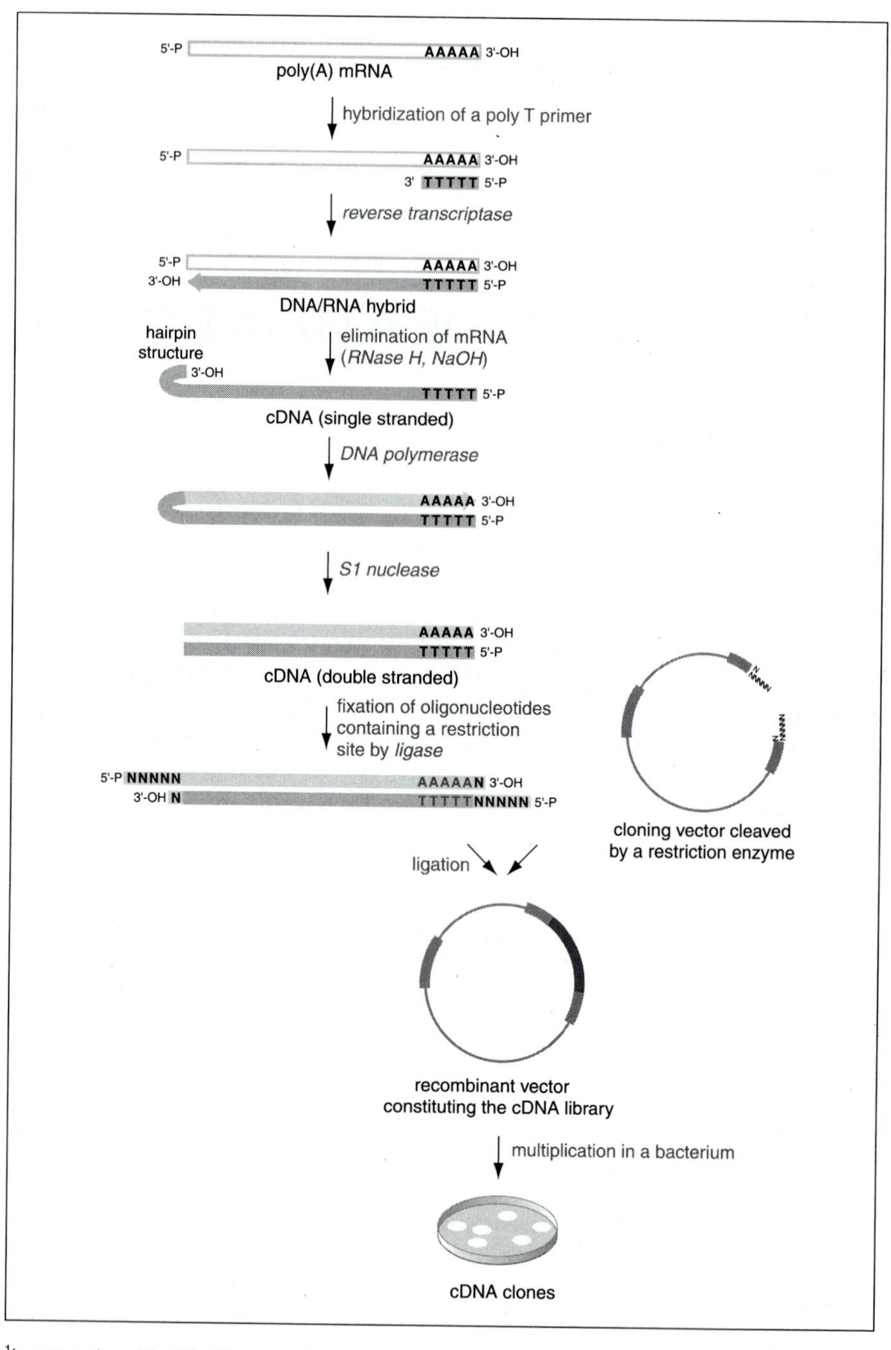

[1]In cases where it is difficult to extract large quantities of mRNA, a cDNA library can be constructed from total RNA. mRNAs represent only a part of cellular RNA, constituted essentially of ribosomal RNA. The PCR technique is particularly suitable for the construction of cDNA libraries from very small quantities of RNA.

Profile 16

SCREENING OF A LIBRARY

Screening a genomic or cDNA library consists of specifically selecting bacterial clones containing the desired gene sequence (see also Profiles 17 and 20).

In order to identify the clones of interest, it is essential first to replicate the library. The bacterial colonies or lysis plates are transferred to a nitrocellulose or nylon membrane (see Profile 2). Their DNA is denatured by an NaOH incubation, fixed covalently on the membrane, and hybridized with different types of labelled nucleic probes in order to detect positive clones responding to the hybridization.

Use of a DNA fragment as probe

Any sequence of DNA fragment (cDNA, PCR fragment, oligonucleotide) known to have a homology with the desired cDNA or gene can be used as a probe for screening the library. The probe DNA fragment is labelled by random priming or by nick translation (see Profile 11). The hybridization membranes containing the vectors of the library are then incubated with this probe and the positive clones (those hybridizing with the probe) are detected and extracted.

Use of synthetic oligonucleotide as probe

If the sequence of the protein from which a gene is to be isolated is known, it is possible to design a corresponding nucleotide sequence by following the rules of the genetic code. Because of the degenerative nature of the genetic code,[1] several nucleotide sequences are obtained. The combination of oligonucleotides (generally about 20 bases) is synthesized, radioactively labelled (see Profile 11), and hybridized with clones fixed on the filter.

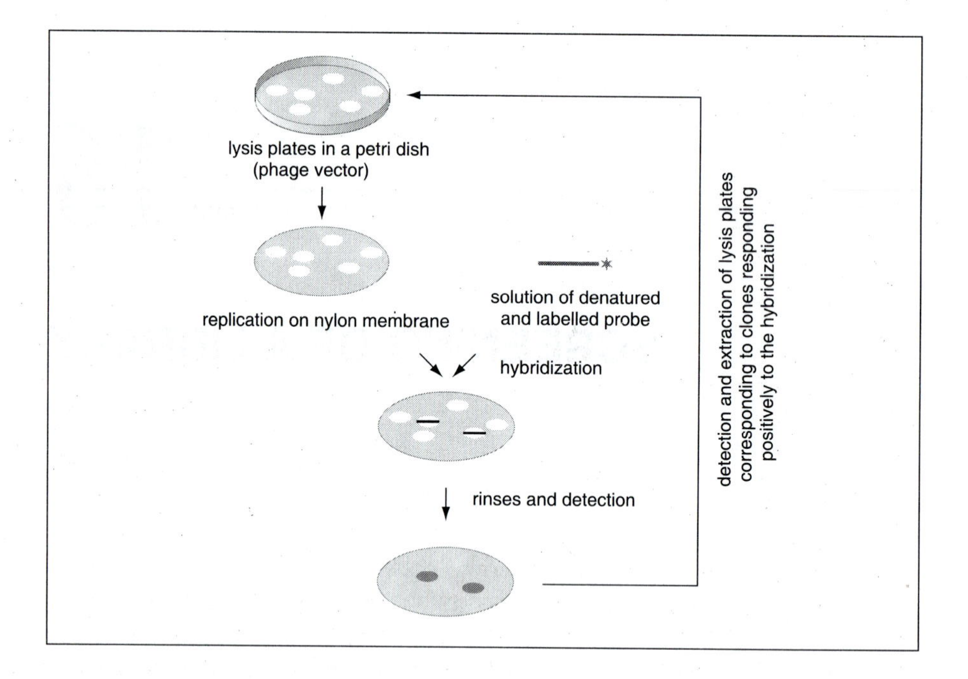

Screening by antibodies

Sometimes nothing is known about the sequence of the protein or the gene to be cloned. A biochemical approach can be taken to obtain the purification of a protein and a specific antibody for this protein. Such antibodies can be used to isolate a corresponding cDNA. This supposes that the cDNA library has been constructed in an expression vector: the insertion site of the cDNA in the vector is found downstream of a bacterial promoter. Thus, all the bacteria of the library containing these recombinant vectors are capable of synthesizing the protein corresponding to the cDNA sequence.[2] The hybridization membranes containing the cDNA library are incubated with the antibodies directed against the protein from which the cDNA is to be cloned. The bacteria that have the corresponding cDNA will be able to synthesize the protein that will be recognized by the antibodies. For this to happen, the cDNA must be in a correct reading frame in order for the code to be read correctly. The stable antigen-antibody complex is revealed by a colorimetric reaction using a second antibody coupled with alkaline phosphatase. The clones responding to this hybridization are thus found in the petri dish and extracted.

Search for transcription factors

When cDNA libraries were obtained in expression vectors, techniques were

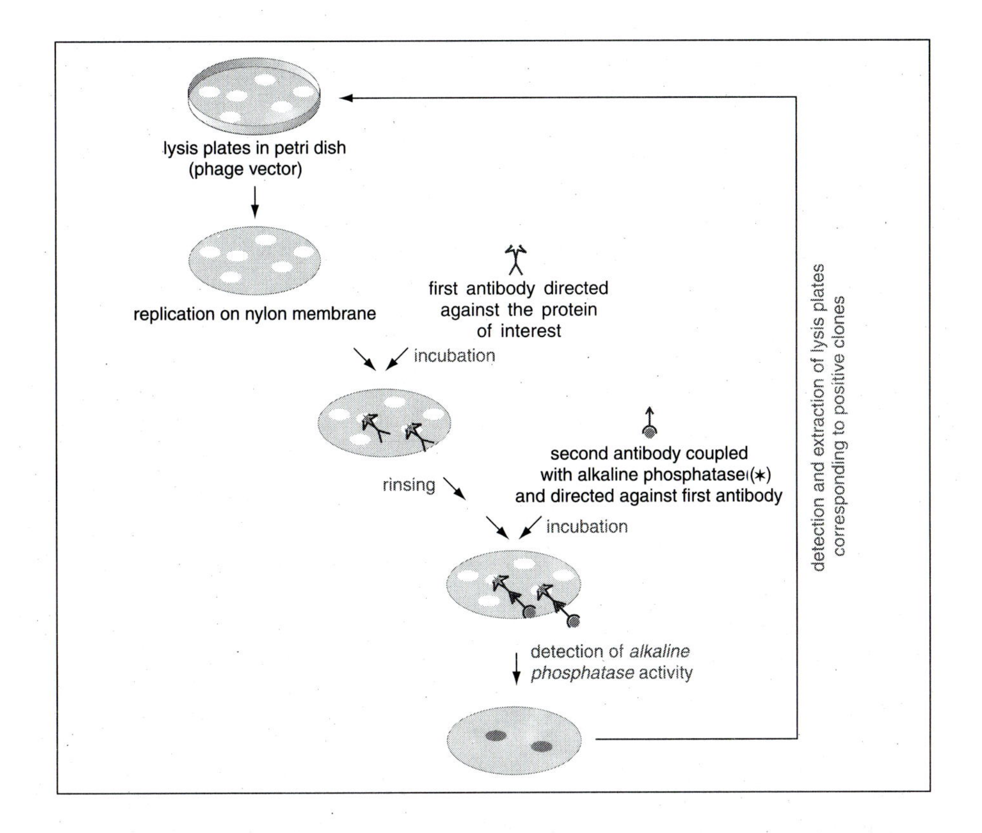

developed to clone genes coding for proteins binding to regulatory DNA sequences. These proteins are factors of transcription regulation. The library of expression is screened with a radioactively labelled DNA fragment that has already been demonstrated to be able to complex with nuclear proteins (see Profiles 31 and 32). If a bacterium from the library contains a recombinant vector expressing a protein binding to this DNA fragment, it will be detected by hybridization. The mRNAs corresponding to proteins of regulation are often accumulated in small quantity in the tissues. The cDNAs of these mRNAs are thus poorly represented in the libraries and a large number of clones (around 10^6) must be screened to obtain a result.

---

[1]An amino acid is coded by several different codons: this is the degeneracy of the genetic code. Example: alanine is coded by the four codons GCU, GCC, GCA, and GCG.

[2]The synthesis of the protein must occur in a correct reading frame. The vectors of expression used are generally phages of the λ type (λ gt11). Since the genetic code is read 3 bases at a time, there are 3 possible reading frames. But the cDNA can be cloned in the right direction (beginning with the coding strand towards the bacterial promoter) or in the wrong direction (beginning with the coding strand away from the promoter). In total, there is 1 chance in 6 that the cDNA will be in a correct reading frame. If the cloning of cDNA is directional (the restriction sites bordering the 5′-P and 3′OH ends of the cDNA are different and allow for a proper orientation of the cDNA with respect to the bacterial promoter), then there is 1 chance in 3 that the cDNA is in a correct reading frame.

Profile 17

DIFFERENTIAL SCREENING: SUBTRACTIVE LIBRARIES, AFLP-cDNA

Differential screening is a technique for cloning cDNAs corresponding to several specific mRNAs of a given tissue at a particular stage of development or under the effect of stress. The principle consists primarily of constructing a cDNA library from mRNA obtained by a treatment in which the specific cDNAs are to be isolated (e.g., leaves). Then, single-stranded (ss) cDNAs of the target tissue and another tissue (e.g., roots) are synthesized by reverse transcription of RNAs extracted from the two types of tissue. These cDNAs are then labelled to be used as probes. The library is then partly screened with the two probes. Some cDNA clones from the library will hybridize with the two probes: these correspond to mRNAs accumulated in the two types of tissue. In contrast, some cDNAs will hybridize only with the probe of the cDNAs from the target tissue: these correspond to specific mRNAs of the target tissue. These clones are then purified and characterized.

This differential screening technique is effective only for isolating the most abundant mRNAs. Several variants have been developed for isolating rare mRNAs, such as subtractive libraries (see also Profiles 18 to 20).

Use of subtractive cDNA library

The presence of abundant mRNAs in a tissue masks the rare mRNAs, which may, however, play a significant role in the mechanisms studied. In order to increase the chances of cloning cDNAs corresponding to the rare mRNAs, the abundant mRNAs can be eliminated before the differential screening. This is called a subtraction of cDNAs. There are several possible approaches. For example, ss cDNAs can be produced from mRNAs of cells of a tissue (such as root) other than the target tissue (leaf). These later cDNAs are

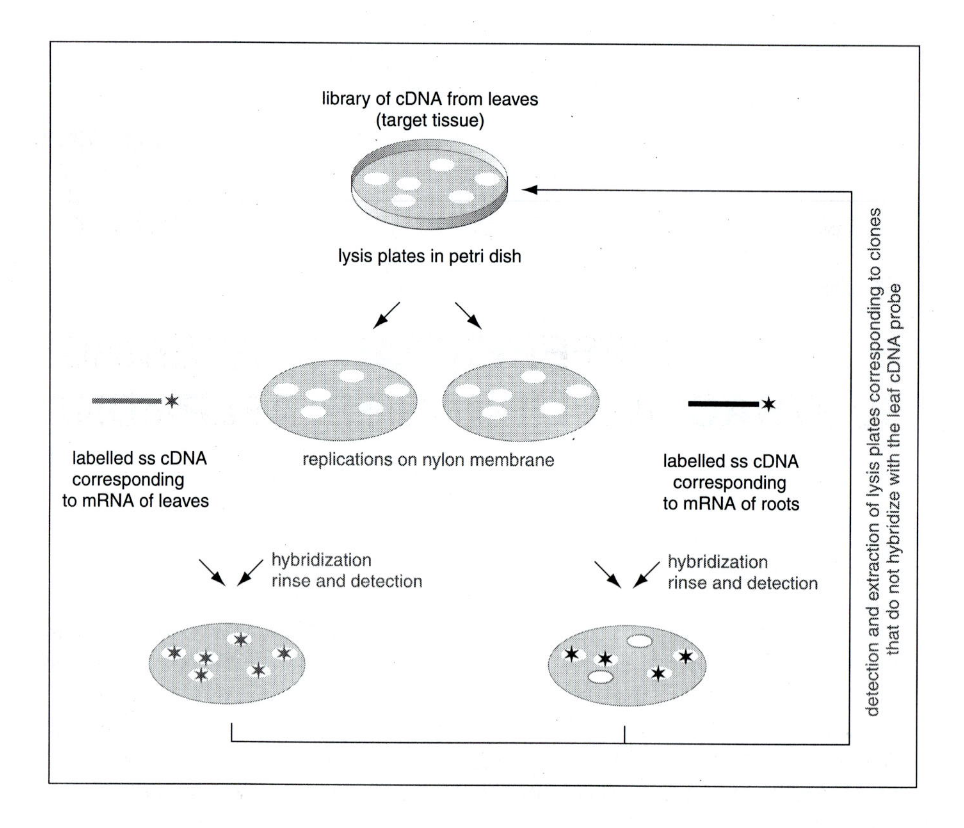

modified, for example by the addition of a nucleotide coupled with biotin during their synthesis. The biotinylated cDNAs are then hybridized with a large surplus of mRNAs extracted from cells of the target tissue (the leaf). All the cDNAs corresponding to the gene expressed in the two cell types will form cDNA-mRNA hybrids. The cDNAs specific to roots as well as those specific to the leaves remain in the single-stranded form. All these RNAs (ss or in the form of cDNA/RNA hybrid) are passed through a column of streptavidin, very similar to biotin. All the RNAs of the non-target tissue, whether ss or hybridized with cDNA, are retained by this column. The eluent is highly rich in mRNAs specific to leaves. These mRNAs then serve as a template to constitute a cDNA library, which is in turn subjected to a conventional differential screening.

RNA screening by cDNA AFLP

The AFLP technique (see Profile 52) can be used to identify coding sequences expressed in a differential manner in two populations of messenger RNAs. One major difference with respect to AFLP on genomic

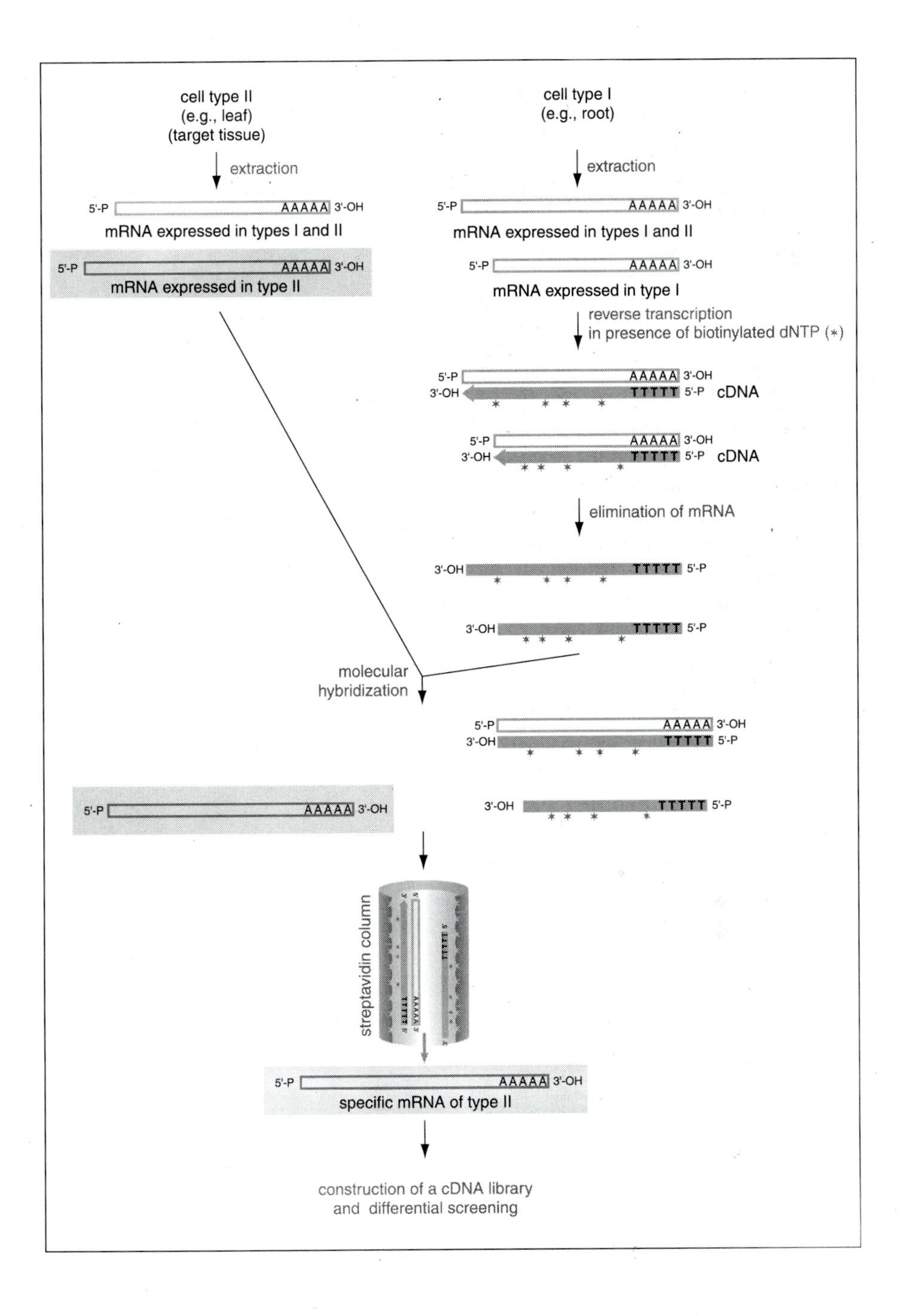

cell type II
(e.g., leaf)
(target tissue)
extraction
5'-P
AAAAA
3'-OH
mRNA expressed in types I and II
mRNA expressed in type II
cell type I
(e.g., root)
mRNA expressed in type I
reverse transcription
in presence of biotinylated dNTP (*)
TTTTT
cDNA
elimination of mRNA
molecular
hybridization
streptavidin column
specific mRNA of type II
construction of a cDNA library
and differential screening

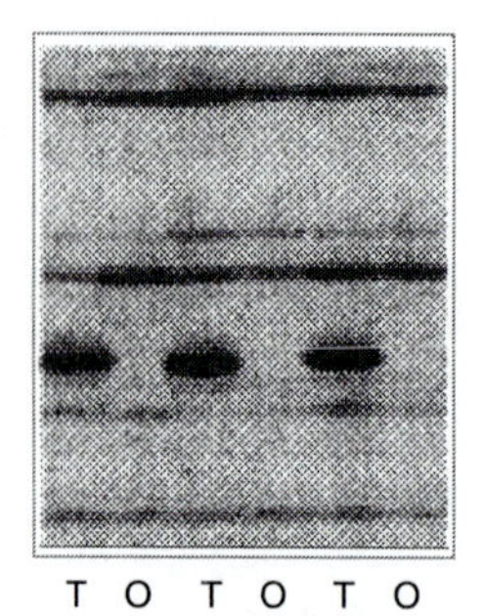

The mRNAs have been isolated from a T tissue and an O tissue.
The cDNAs have been synthesized and digested by *Eco* RI and *Mse* I.
Pre-amplification: primers without extension (primer +0)
Amplification: primers with extension of 2 bases (primer +2)
a, b and c represent 3 repetitions

DNA arises from the fact that a population of cDNAs has a narrower variability of sequences (only the coding sequences are represented), but a highly variable level of representation of these sequences (linked to the expression of the genes). Because of this, the amplification of restriction fragments is generally carried out in much less selective conditions. Often, the pre-amplification is not selective (primers without extension) and the selective amplification is obtained using primers carrying two extensions (primers +2) (see Profile 52).

Profile **18**

DIFFERENTIAL DISPLAY RT-PCR

The technique of differential display RT-PCR is a variant of differential screening adapted to the identification of rare mRNA (see Profile 17). It consists of three major steps: (1) reverse transcription from mRNAs resulting in the obtaining of single-stranded cDNAs (see Profile 27); (2) PCR amplification of the cDNAs; (3) electrophoretic separation of fragments of double-stranded cDNAs before they are cloned. These three steps are carried out simultaneously and in parallel on mRNAs isolated from different tissues or treatments. The electrophoretic gels can be compared to identify bands of cDNA that are differentially accumulated.

Reverse transcription

After extraction of total RNAs and possibly a selection of poly(A) mRNAs (in the figure from cells A and B, which can be compared), the first step is to synthesize single-stranded cDNAs. This reverse transcription is effected using a primer of the $T_{12}XY$ type, complementary to the poly(A) tail. X can correspond to any base except thymidine and Y to any base. Thus, there are 12 possible combinations of $T_{12}XY$ primers, all complementary to different mRNA families. For example, the $T_{12}AC$ primer will be complementary to all the mRNAs ending in GUA_{12}, while the $T_{12}AG$ primer will be complementary to all the mRNAs ending in CUA_{12}. It is the choice of primers that determines the screening of the mRNAs analysed. This screening is necessary to improve the resolution of the gel electrophoresis subsequently. If no screening is used (use of a T_{12} primer that is complementary to all the mRNAs), the quantity of cDNAs obtained will be too large and the resolution of the gel electrophoresis will be poor, since too many bands will appear. This screening step makes it possible to select a population corresponding on average to one twelfth of the population of poly(A) RNAs present in the cell type studied. When a primer is chosen, it is the same one used on the mRNAs isolated from the different tissues or treatments studied.

PCR amplification

During this step, the single-stranded cDNAs obtained by reverse transcription are amplified by PCR (see Profile 24). The amplification requires two primers: the first, complementary to the 3'-OH end of the cDNAs, will be the $T_{12}XY$ primer used during the reverse transcription and the second, complementary to the 5'-P end of the cDNAs, will correspond to a random primer of 10 base pairs (see Profile 51). The random primer hybridizes at random in the 5'-P part and, in relation to the poly(A) tail, at a certain distance that varies for each cDNA selected during the first step. This leads to the obtaining of double-stranded partial cDNAs of different sizes, which enclose the 3'-OH part of the corresponding mRNAs.

Electrophoretic separation and cloning of cDNAs

The double-stranded cDNAs derived from the PCR amplification step are separated on acrylamide gels similar to those used for sequencing. These gels generally resolve at the base for fragments of less than 500 bp, the type of bands obtained by this method. The various products are visualized either by autoradiography (if the preceding step of PCR was carried out in the pre-sence of [^{35}S] dATP) or by silver staining. Different samples must be compared on the same gel. If the mRNAs are differentially accumulated between two treatments, in a reproducible fashion, it is possible to excise differentially accumulated bands. This can be done directly when the gel is detected by silver staining. In contrast, with respect to radioactive labelling, the autoradiogram of the gel must be superimposed in order to delimit the zone containing the target cDNA band. The quantity of cDNA present in a band is too small to be directly sub-cloned. At this stage, it is necessary to reamplify the DNA present in the excised band using the primer pair used earlier. The new fragment obtained is then sub-cloned in the plasmid (see Profile 9) and sequenced (see Profile 23). A probe generated from this sequence will be used for a more precise study of the expression of the corresponding gene: screening of a cDNA library (in order to obtain a full-length cDNA) or a genomic library, northern, *in situ* hybridization, etc.

There are many advantages to the technique of DD-RT-PCR:

- Based on the use of PCR, the technique requires little material: 1 µg of poly(A) RNAs can be used for 150 reactions.
- Over- and underexpressed genes can be identified on the same gel (bands appearing or disappearing).

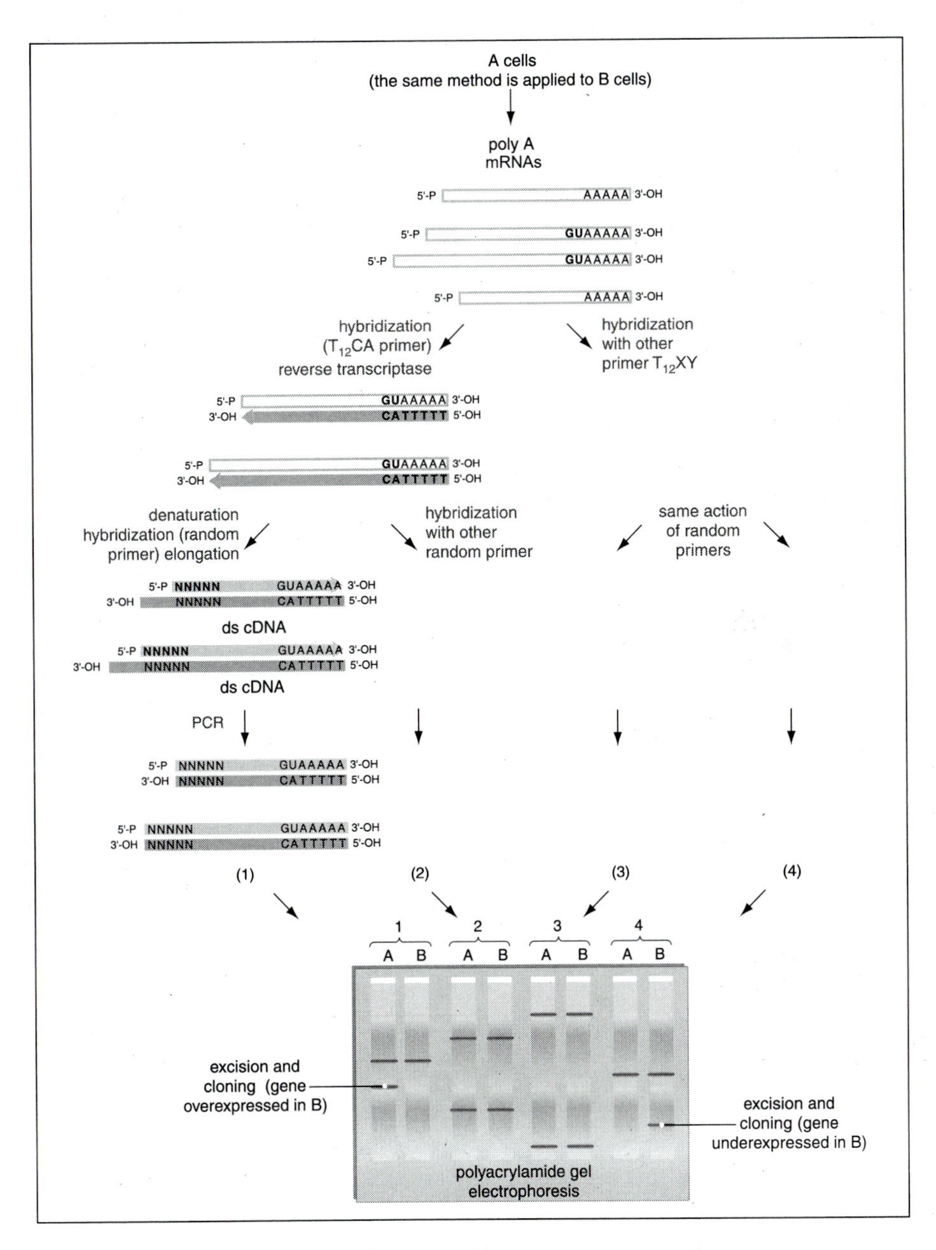

- It is much faster than conventional techniques of differential screening and DD RT-PCR can be used to isolate and identify a cDNA clone in 15 days.

An approach similar to DD RT-PCR can be followed using AFLP (see Profile 17).

JEAN-LOUIS HILBERT

Profile **19**

DIFFERENTIAL SCREENING BY SSH: SUPPRESSION SUBTRACTIVE HYBRIDIZATION

Suppression Subtractive Hybridization (SSH) is a differential screening method used along with PCR. The objective is to suppress the mRNAs common to two samples or two cell types in order to create simplified sub-populations containing only the mRNAs expressed differentially between the initial samples. Very similar to RDA (see Profile 20), SSH is distinguished essentially by a standardization step between the abundant sequences and those that are very rare, which is designed to favour the detection of mRNAs that are less abundant. It is based simultaneously on the kinetics of reassociation of DNA strands (abundant fragments will hybridize more quickly than rare fragments) and on the existence of the suppressor effect of PCR. This effect is due to the fact that long reverse (complementary) adaptors placed on either side of a single strand will have greater affinity for each other than for a shorter amplification primer. When these long reverse adaptors recognize each other, they form non-amplifiable hairpin structures.

As in most subtraction techniques, the term *tester* is used for the target sample and *driver* for the control sample to which it is compared. The RNAs

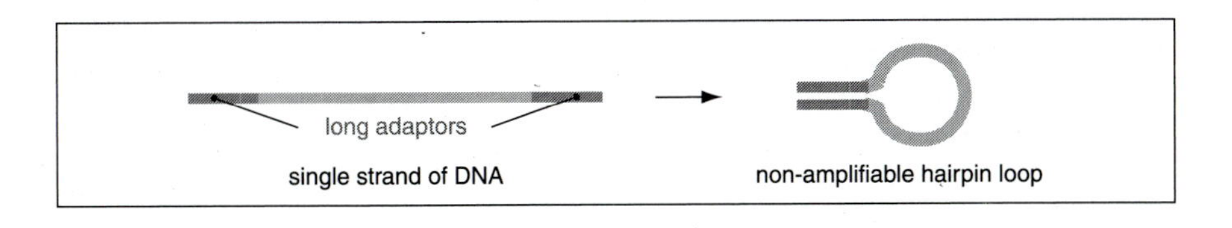

are extracted from two types of tissues, then transformed into single-stranded cDNAs by reverse transcription and into double-stranded cDNAs by DNA polymerization. The tester and driver cDNAs are first cleaved by a single enzyme in order to obtain small fragments, which are more favourable to the subsequent steps of the hybridization. The tester cDNAs thus cleaved are divided into two equal pools (tester1 and tester2) to which different long adaptors (44 bp) are linked. No adaptor is ligated to cleaved double-stranded driver cDNA.

The three lots of cDNA (driver, tester1, and tester2) are denatured. The surplus driver cDNAs are combined with each of the tester lots. The cDNA combination is left to reassociate. There are several outcomes: in the solution of each lot, single-stranded (ss) sequences will be found that have not yet found their complementary strand, as will tester/tester or driver/driver homohybrids (standardizing effect of the kinetics of hybridization) and tester/driver heterohybrids (subtractive effect of the hybridization). These homo- or heterohybrid molecules can no longer participate after the reassociation.

The two tester1 and tester2 lots are then reunited and the reassociation is pursued for 16 to 18 h, often in the presence of a new driver surplus. Ultimately, the solution contains tester1/tester2 heterohybrids, in addition to the homohybrids and heterohybrids mentioned earlier. These heterohybrids correspond to sequences that are absent from driver and that are thus specific to tester. These are the target sequences of SSH.

After the 3'-OH ends are filled up with regard to adaptors, a couple of primers complementary to the two types of adaptors are used to carry out a PCR amplification of the target sequences. The sequences without adaptor (driver/driver homohybrids) will not be amplified, the sequences possessing an adaptor at only one end (tester/driver heterohybrid) will be amplified but linearly, and the sequences possessing a single adaptor at each end (tester1/tester1 or tester2/tester2 homohybrids) will fall under the suppressor effect of the PCR because of the presence of complementary sequences at the ends of each strand. Only the tester1/tester2 heterohybrids will be amplified exponentially.

In a single subtraction cycle, the SSH allows the creation of a subpopulation of cDNAs that can be cloned or directly observed on an agarose gel after electrophoresis. The product or products of the most intensive amplification can thus be easily recovered to be sequenced or used as probes to screen a genomic or cDNA library. The standardization step makes it possible to avoid the biases due to differences in the abundance of mRNA, a factor to which RDA and most RNA screening techniques are sensitive. Since the enrichment factor ranges from 1000 to 5000, SSH is also a method of choice to find and clone rare sequences. The implementation of SSH can also be limited in some cases by the initial quantity of material

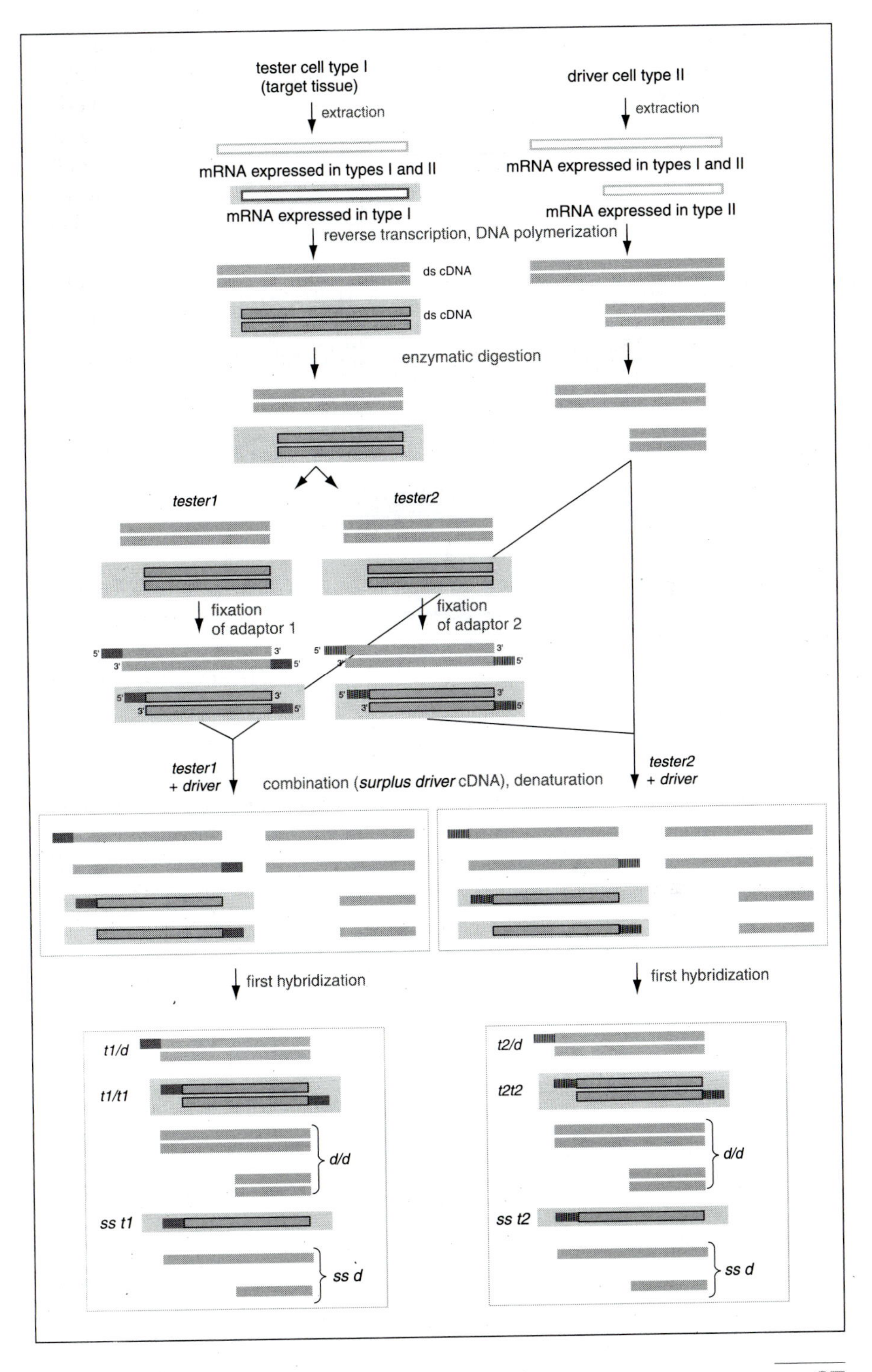
tester cell type I
(target tissue)
driver cell type II
extraction
extraction
mRNA expressed in types I and II
mRNA expressed in types I and II
mRNA expressed in type I
mRNA expressed in type II
reverse transcription, DNA polymerization
ds cDNA
ds cDNA
enzymatic digestion
tester1
tester2
fixation
of adaptor 1
fixation
of adaptor 2
tester1
+ driver
combination (surplus driver cDNA), denaturation
tester2
+ driver
first hybridization
first hybridization
t1/d
t1/t1
d/d
ss t1
ss d
t2/d
t2t2
d/d
ss t2
ss d

needed: this technique requires at least 1 µg of poly(A) RNA. Moreover, the cDNA restriction step hinders the characterization of entire (full-length) cDNA sequences.

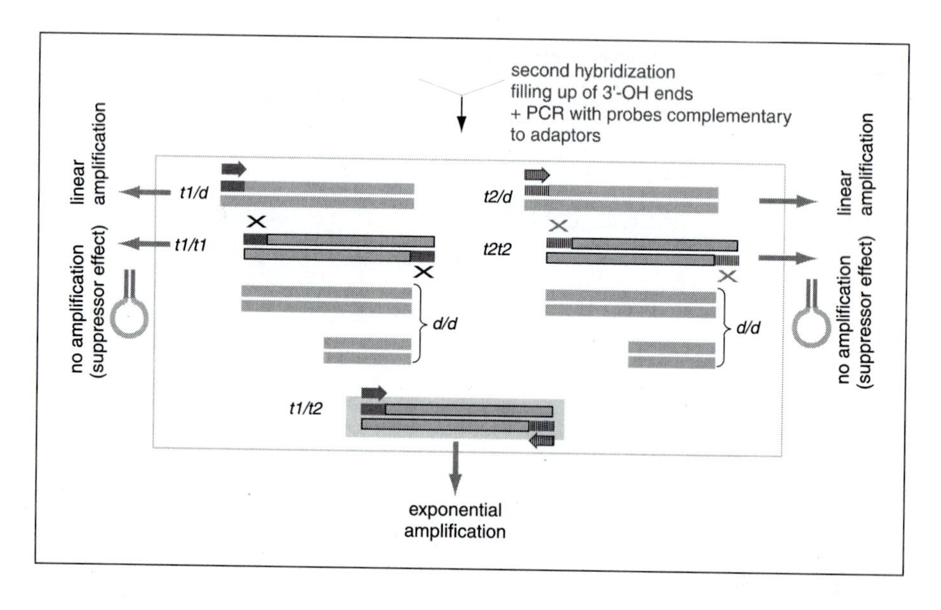

Profile **20**

DIFFERENTIAL SCREENING BY RDA: REPRESENTATIONAL DIFFERENCE ANALYSIS

The subtraction of cDNA by RDA is a differential screening technique assisted by PCR, which allows the identification of genes expressed specifically in a target tissue in relation to a control tissue. The subtraction comprises the following steps: (1) synthesis of double-stranded cDNAs and their cleavage by a restriction enzyme; (2) their multiplication by PCR amplification; (3) subtractive hybridization. Several cycles of subtractive hybridization are effected so as to enrich the product obtained in differentially expressed genes.

Synthesis of double-stranded cDNAs and digestion

RNA molecules are extracted from the target tissue (or tester) and control tissue (or driver). The poly(A) mRNAs are selected and the corresponding double-stranded (ds) cDNAs are synthesized. These cDNAs are then cleaved by the restriction enzyme *Dpn* II, which produces protruding sticky 5'-P ends and has a recognition site 4 bp long.

PCR multiplication

Two synthetic oligonucleotides are hybridized at the 5'-P and 3'-OH ends of cDNA fragments from two tissues, so as to produce adaptors after ligation. The oligonucleotide R12 has a part complementary to the sticky end of site *Dpn* II and will place itself on this DNA strand. However, because of the absence of a phosphate group at its 5'-P end, it cannot form a covalent bond with the cDNA strand. In fact, it will serve as a guide for a second oligonucleotide, R24, which is partly complementary to it. This

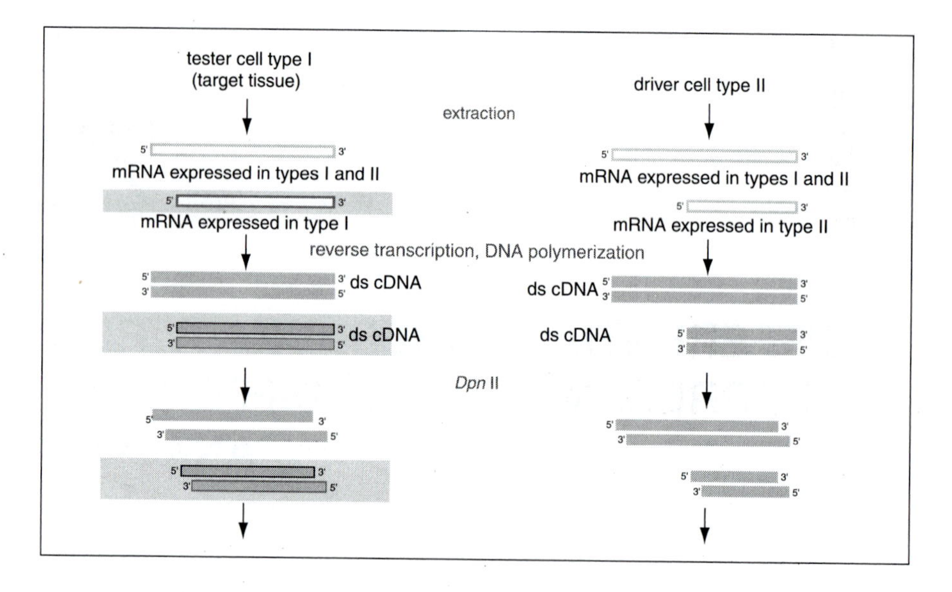

oligonucleotide R24 fixes covalently on the cDNA. By this means, populations of cDNAs are obtained that have the adaptor R24 at their 5'-P ends. After elimination of the adaptor R12 and filling up of the 3'-OH ends, the cDNA fragments are amplified by PCR in the presence of primers that hybridize with the R24 adaptor.

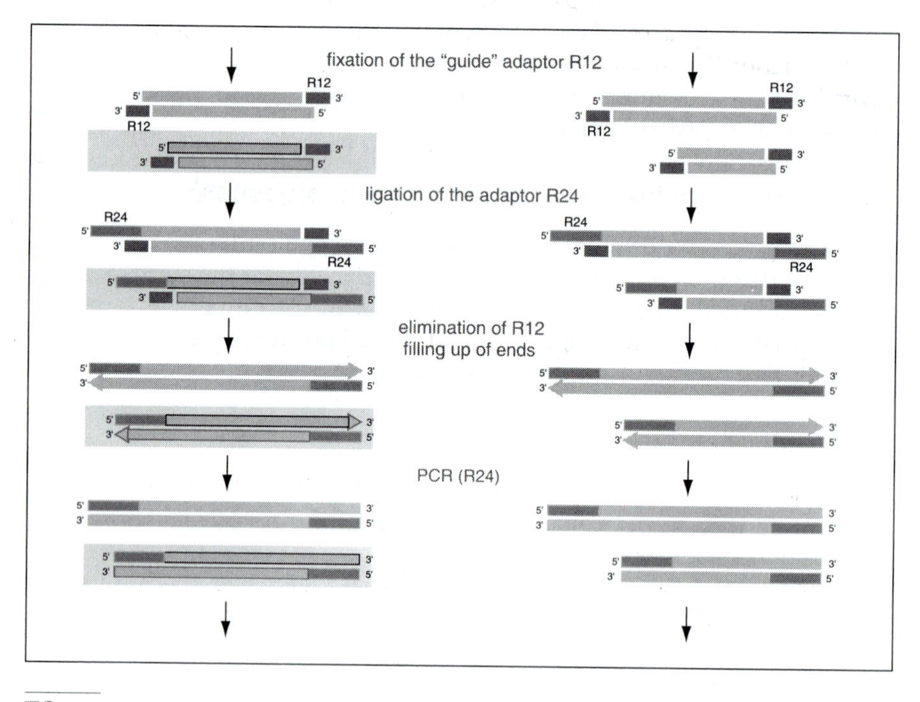

Subtractive hybridization

The amplified cDNA fragments from two tissues are again cleaved by the enzyme *Dpn* II in order to eliminate the adaptor R24. New adaptors (J) are ligated at the 5'-P ends of the fragments of cDNAs of target tissues only. The cDNAs of tester and driver tissues are combined in a ratio of 1:100. The mixture is heated so as to dehybridize the fragments of double-stranded cDNA, then cooled so that the strands can hybridize again. Since there is a combination of cDNAs from two tissues (one of which is surplus), the single-stranded cDNAs of the target tissue can hybridize with their complementary strand of the tester tissue, but mostly with those of the driver tissue, which are surplus, except for the cDNAs present only in the tester tissue, i.e., the cDNAs that are of interest. After filling up of the 3'-OH ends with respect to the adaptor, a PCR amplification is carried out and three possible outcomes may result: (1) when the gene is expressed only in the tester tissue, the duplex will comprise an adaptor J on each strand and the PCR will produce an exponential amplification; (2) when the gene is expressed in both tissues, the duplex comprises only a single strand possessing an adaptor J that undergoes a linear amplification; (3) when the gene is expressed only in the driver tissue, the duplex formed has no adaptor J and is not amplified. The PCR products are cleaved by the exonuclease *Mung Bean Nuclease* specific to the single-stranded DNA, so as to eliminate the products of linear amplification. The restriction products are again amplified by PCR using J primers.

So that a second subtractive hybridization can be carried out on the PCR products obtained, the J primers are eliminated by the *Dpn* II enzyme and new adaptors are ligated at the 5'-P end. A new subtractive hybridization identical to the first is applied so as to increase the product of the RDA in genes specifically expressed in the tester tissue. This increase can be pursued as many times as necessary by successive subtractive hybridizations. Subsequently, the population of cDNAs enriched in specific sequences of tester tissue can be cloned in plasmidic vectors for conservation and analysis.

This technique presents the same advantages as suppression subtractive hybridization (SSH, see Profile 19), but it is known to generate much fewer false positives than the latter. On the other hand, no commercial kit is available and RDA takes longer to carry out than SSH.

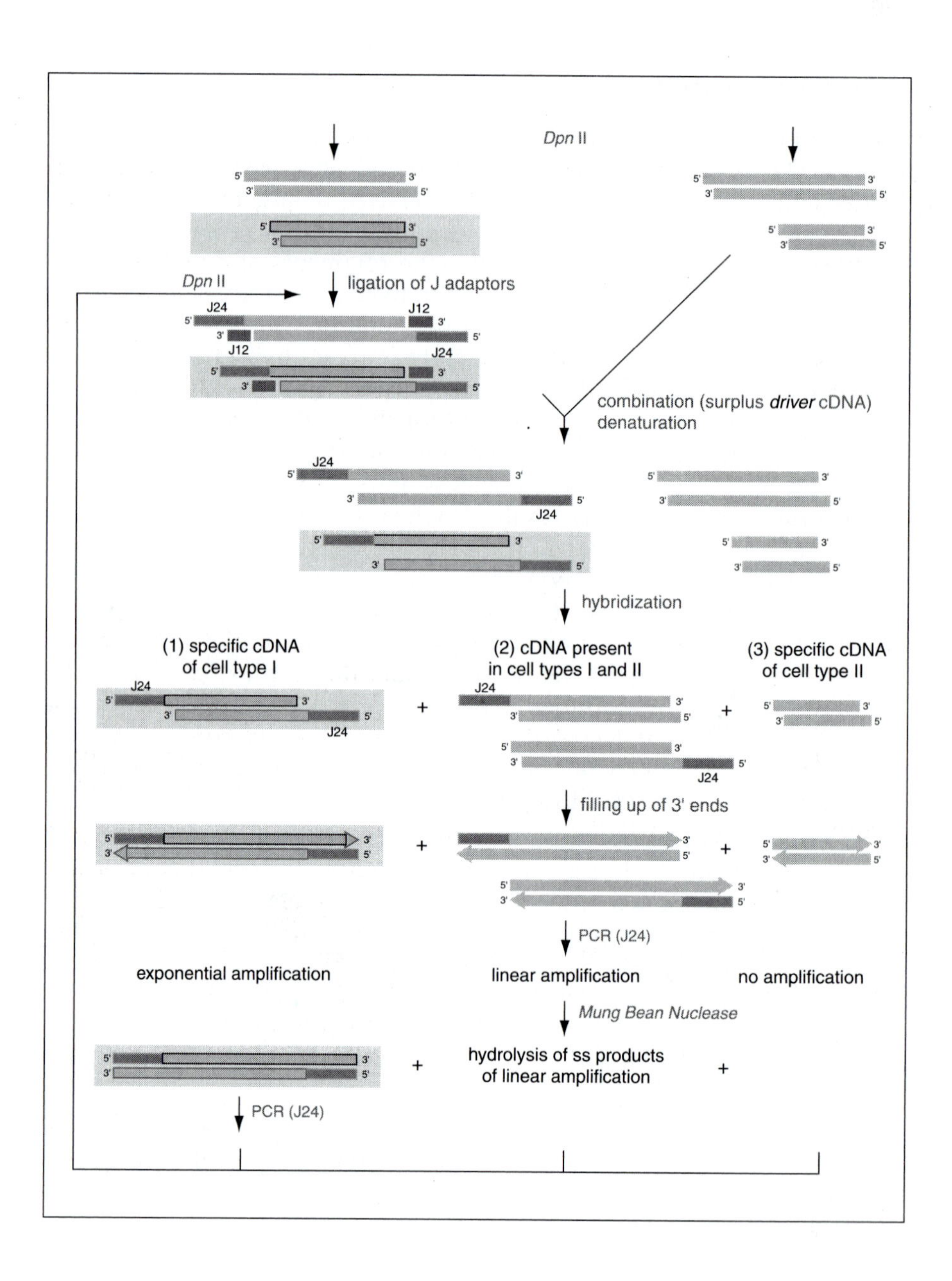
Dpn II
Dpn II
ligation of J adaptors
J24
J12
J12
J24
combination (surplus driver cDNA)
denaturation
J24
J24
hybridization
(1) specific cDNA
of cell type I
(2) cDNA present
in cell types I and II
(3) specific cDNA
of cell type II
J24
J24
J24
J24
filling up of 3' ends
PCR (J24)
exponential amplification
linear amplification
no amplification
Mung Bean Nuclease
hydrolysis of ss products
of linear amplification
PCR (J24)

Profile **21**

EST: EXPRESSED SEQUENCE TAGS

The genome of a eukaryotic organism has many genes but each type of cell expresses only a fraction depending on the functions served. As has been seen earlier (Profile 15), a cDNA library obtained from a particular tissue provides a "snapshot" of the genes expressed in that tissue at the time of extraction of mRNAs. Theoretically, the thousands of phage or plasmid clones of the library cover almost all the sequences expressed in the form of mRNA. The strongly expressed genes are represented more than weakly expressed genes (rare mRNAs).

Expressed sequence tags (EST) are designed to characterize part of the sequence of most of the mRNAs accumulated in a given tissue. They involve the partial sequencing of each of the cDNAs present in a cDNA library. These sequences are then compared one by one to those that are stored in international data banks (e.g., GenBank via Internet). The information obtained on the homology of sequences is in some cases sufficient to assign a potential function to the corresponding gene. It can be hypothesized that ESTs that do not have homology with known genes correspond to new genes specific to the system studied (species, tissue, or physiological state). This search for exhaustivity (to sequence all the expressed genes) is achieved by favouring quantity over quality. However, the information obtained can usually be used to identify the genes and their level of expression in the tissue and the physiological state considered.

The EST approach thus allows the elaboration of a catalogue of genes expressed in a particular tissue. It can be coupled with large-scale studies of expression: each cDNA identified is deposited on a membrane or a glass slide to constitute high-density arrays comprising hundreds, even thousands of cDNAs (see Profile 22). These membranes or slides can then be hybridized with probes corresponding to mRNA of any other tissue or cell subjected to various biotic or abiotic stresses: it is thus possible to study simultaneously the expression of hundreds of genes deposited on filters.

ESTs are also used for the annotation of genomes (location of exons) and in the search for polymorphism for genetic studies.

Method

An EST project comprises a molecular biology step and a bioinformatic step.

- Steps of molecular biology: (1) extraction of RNAs and facultative selection of the poly(A) fraction; (2) construction of the cDNA library (see Profile 15); (3) mass sequencing without previous screening and without attempting to correct errors.

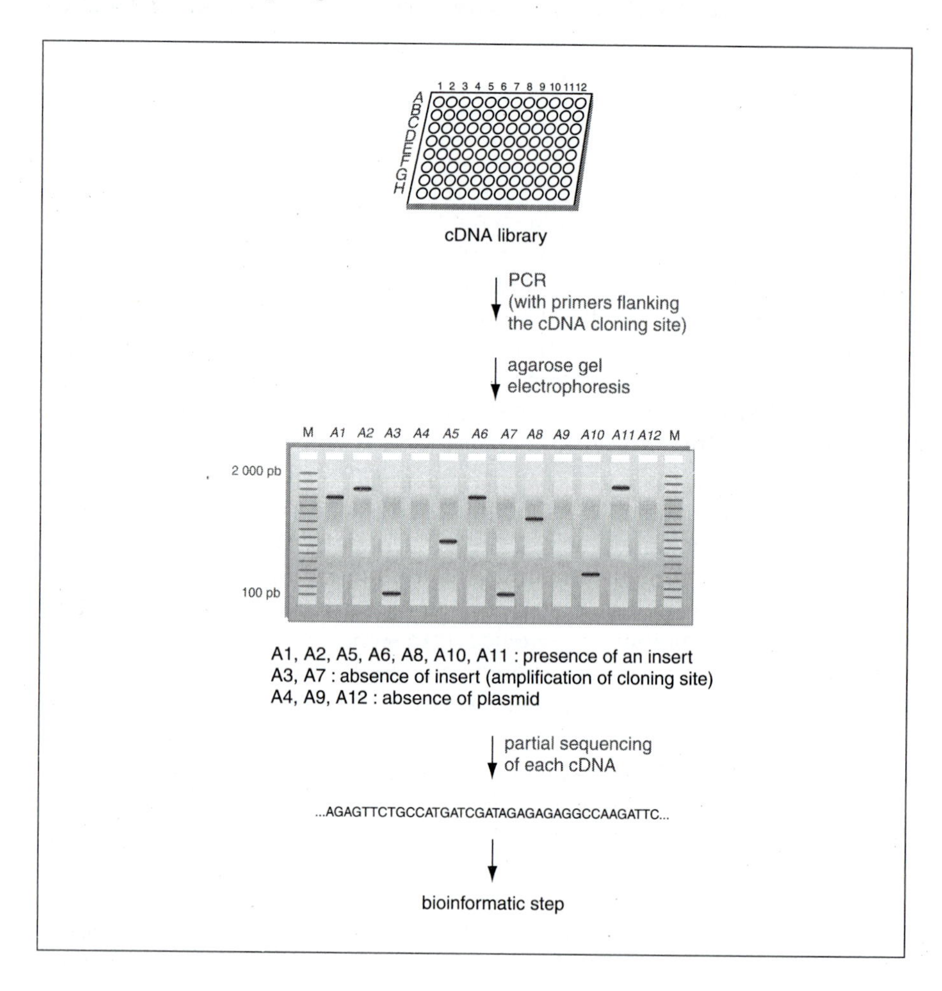

- Bioinformatic steps (analysis of sequence data):
  - Creation of contigs: (1) elimination of zones of insufficient quality; (2) elimination of sequences of vectors and adaptors; (3) elimination of ESTs corresponding to contaminants (ribosomal RNA, bacterial sequences, mitochondrial sequences, etc.); (4) masking of repeat sequences; (5) clusterization, which consists of grouping the ESTs coding for the same gene, on the basis of a strong similarity of sequence; (6) joining, consisting of creating a "consensus" sequence for each gene from several ESTs covering different parts of the gene, i.e., the longest possible contig from overlapping fragments of the cluster.
  - Annotation: this step consists of assigning a function and characterizing the contigs and isolated ESTs (singletons). It is based on investigation of databases of sequences (nucleic and protein), motifs, mass spectrometry profiles of proteins, etc. The use of predictive methods (of function, cellular location, etc.) may prove useful during this step.

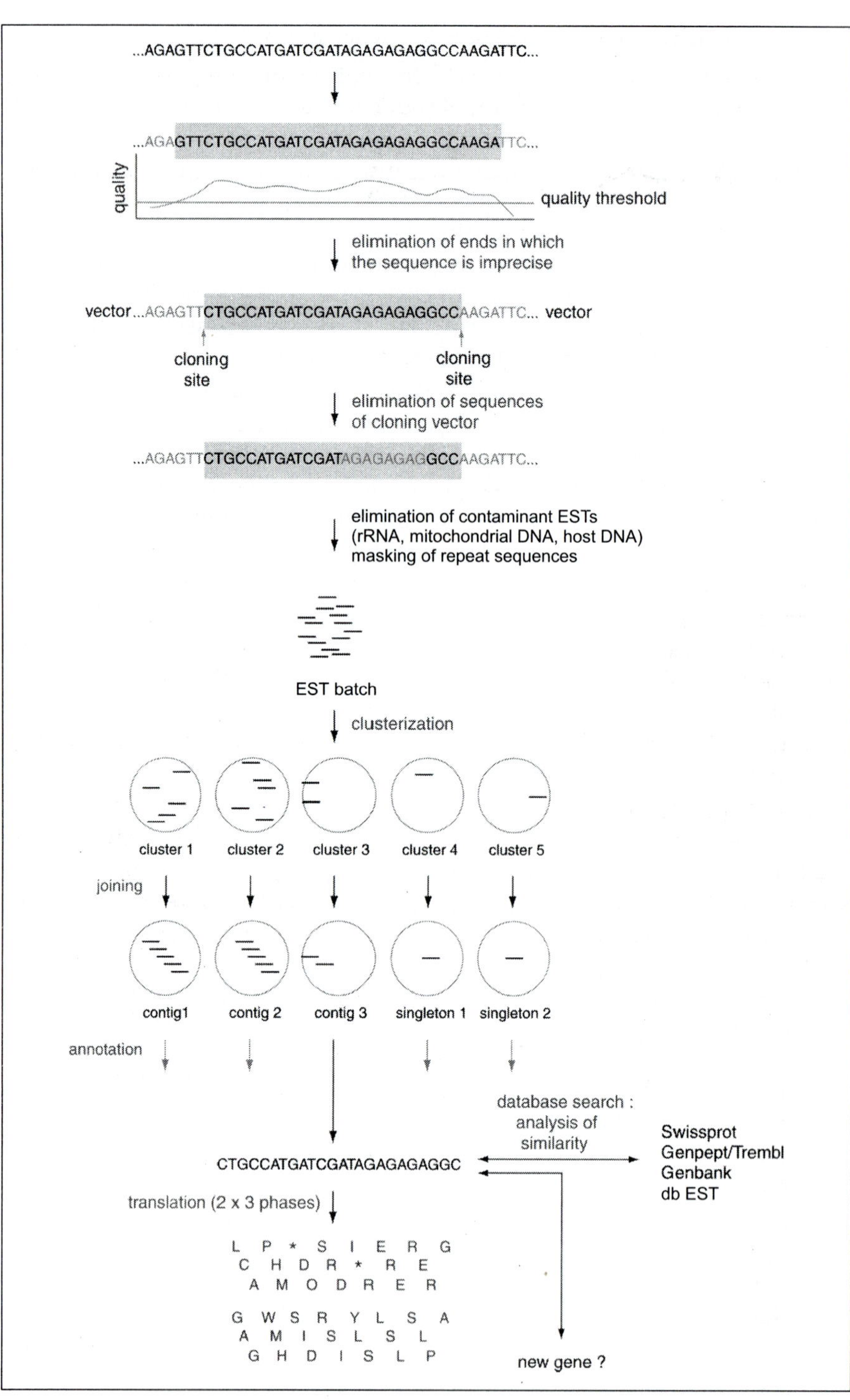

...AGAGTTCTGCCATGATCGATAGAGAGAGGCCAAGATTC...
...AGAGTTCTGCCATGATCGATAGAGAGAGGCCAAGATTC...
quality
quality threshold
elimination of ends in which
the sequence is imprecise
vector...AGAGTTCTGCCATGATCGATAGAGAGAGGCCAAGATTC... vector
cloning
site
cloning
site
elimination of sequences
of cloning vector
...AGAGTTCTGCCATGATCGATAGAGAGAGGCCAAGATTC...
elimination of contaminant ESTs
(rRNA, mitochondrial DNA, host DNA)
masking of repeat sequences
EST batch
clusterization
cluster 1
cluster 2
cluster 3
cluster 4
cluster 5
joining
contig1
contig 2
contig 3
singleton 1
singleton 2
annotation
database search :
analysis of
similarity
Swissprot
Genpept/Trembl
Genbank
db EST
CTGCCATGATCGATAGAGAGAGGC
translation (2 x 3 phases)
L P * S I E R G
C H D R * R E
A M O D R E R
G W S R Y L S A
A M I S L S L
G H D I S L P
new gene ?

Profile **22**

DNA MICROARRAYS: DNA MICROCHIPS, cDNA FILTERS

Some years ago, the analysis of genic expression was possible only for a single gene or a small number of genes at a time (northern blot, RT-PCR). The technique of DNA microarrays[1] now allows the simultaneous measurement of the level of expression of several thousands of genes, or even an entire genome. This technology was developed in the mid-1990s and is now a standard tool for research in molecular biology and clinical diagnostics.

The technology of DNA microarrays is multidisciplinary, being associated with micro-electronics, chemistry of nucleic acids, image analysis, and bioinformatics. It consists of the production and use of arrays containing thousands of samples of nucleic acids attached to solid substrates such as glass microscope slides or nylon filters. Since the area occupied by each sample is small (for glass slides, generally 50 to 200 μm diameter), samples of nucleic acids representing whole genomes, around 3000 to 30,000 genes, can be deposited on a single surface (glass slide or filter). These microchips can be used as probes[2] in experiments of molecular hybridization with targets[2] of labelled nucleic acids. Thus, DNA microarrays can be used to measure the expression of all the genes of a genome simultaneously. This has fundamentally changed the manner in which genome expression is studied: it is called transcriptome analysis.

The term *DNA microarray* can be applied to various types of technologies, each different in the type of nucleic acid fixed on the support and the method of fixation of these molecules. There are two types of microchips: complementary DNA (or cDNA) filters and arrays of oligonucleotides. These arrays are elaborated either by the synthesis of oligonucleotides directly on a glass slide by a photolithographic process or by the deposit (mechanical or electrochemical spotting) of previously synthesized

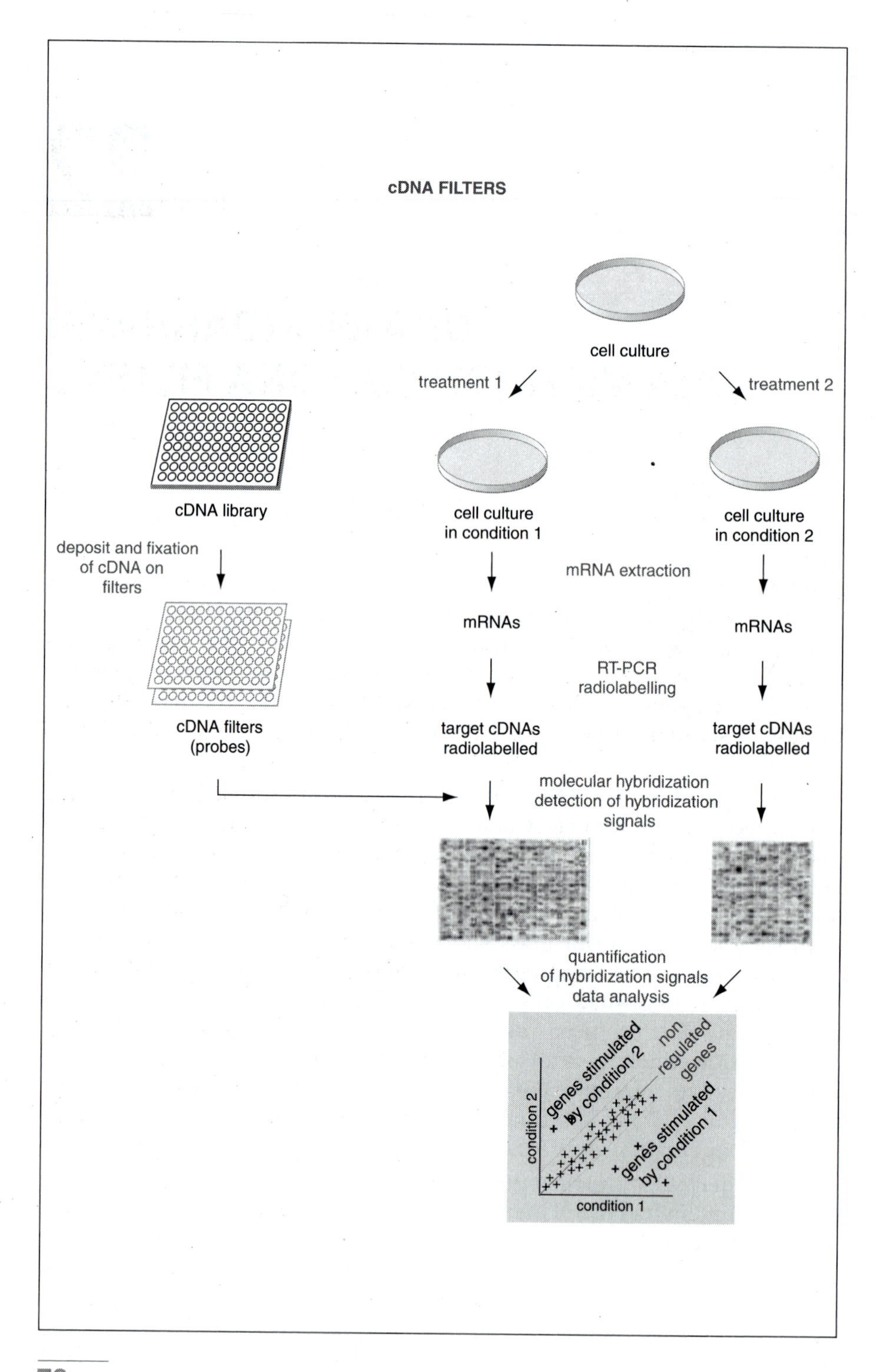
cDNA FILTERS
cell culture
treatment 1
treatment 2
cDNA library
cell culture
in condition 1
cell culture
in condition 2
deposit and fixation
of cDNA on
filters
mRNA extraction
mRNAs
mRNAs
RT-PCR
radiolabelling
cDNA filters
(probes)
target cDNAs
radiolabelled
target cDNAs
radiolabelled
molecular hybridization
detection of hybridization
signals
quantification
of hybridization signals
data analysis
genes stimulated
by condition 2
non
regulated
genes
genes stimulated
by condition 1
condition 2
condition 1

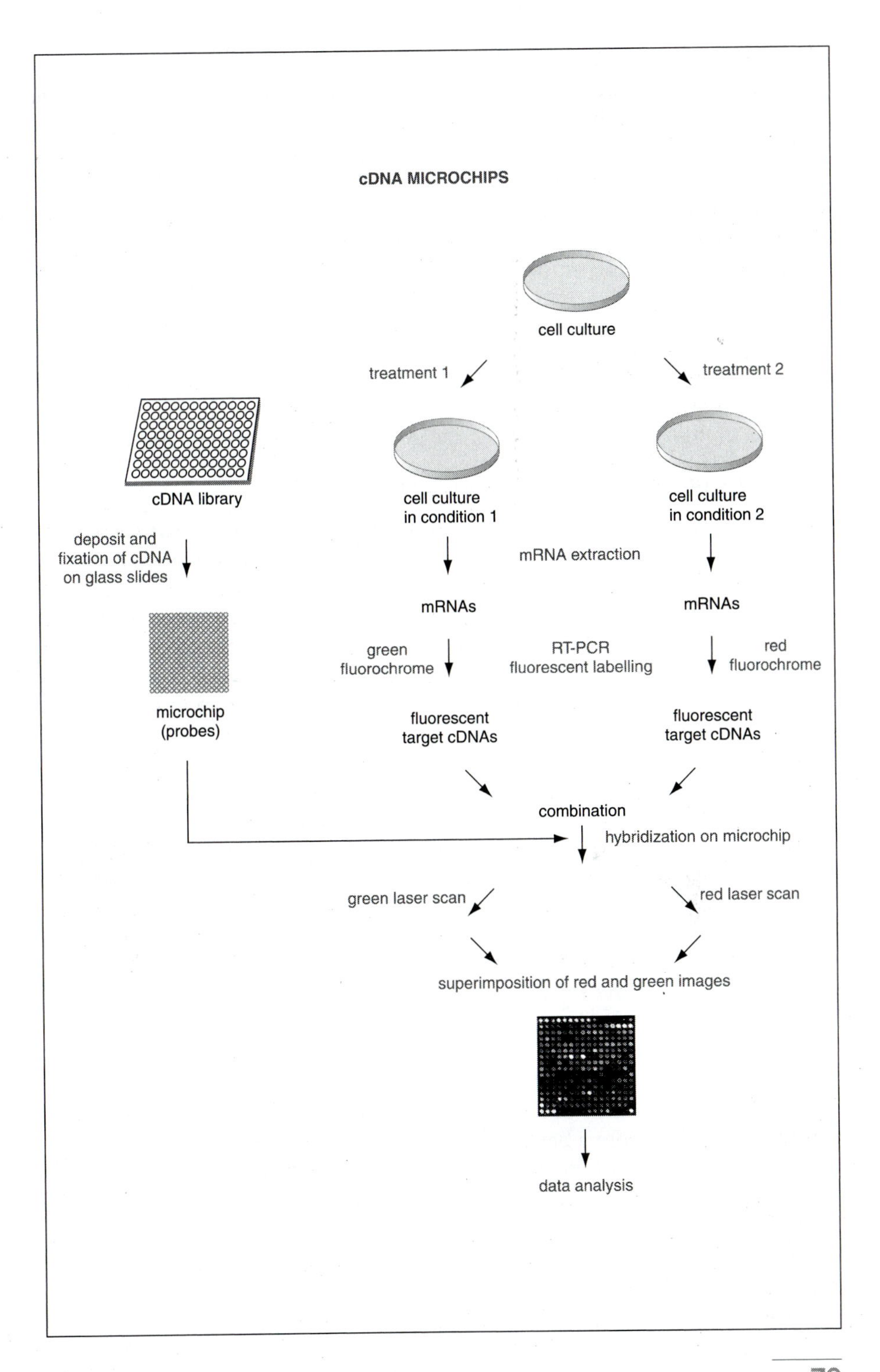
cDNA MICROCHIPS
cell culture
treatment 1
treatment 2
cDNA library
cell culture
in condition 1
cell culture
in condition 2
deposit and
fixation of cDNA
on glass slides
mRNA extraction
mRNAs
mRNAs
green
fluorochrome
RT-PCR
fluorescent labelling
red
fluorochrome
microchip
(probes)
fluorescent
target cDNAs
fluorescent
target cDNAs
combination
hybridization on microchip
green laser scan
red laser scan
superimposition of red and green images
data analysis

oligonucleotides or cDNA on the support (nylon membrane or glass slide). An ideal support would allow an effective immobilization of probes on the surface and a strong hybridization of the probe with the target.

In any case, experiments using DNA microarrays to analyse a transcriptome correspond to comparative hybridizations involving populations of RNAs accumulated in different tissues, at different stages of development, or subjected to different treatments. The various steps of these approaches are described in the figures on pp. 78 and 79.

Elaboration of filters and glass slides

In the case of filters, it is the cDNAs that are deposited. On the glass slides, it is possible to deposit cDNAs or to synthesize directly on the slides oligonucleotides representative of the genes of the organism studied.

Synthesis of oligonucleotides on glass slides

Some DNA microarrays are produced by *in situ* synthesis of oligonucleotides by photolithography. A step-by-step chemical synthesis of oligonucleotides is carried out on the slide. The activated slide (treated so as to fix a nucleotide) is soaked in a solution containing one of the constituent bases of DNA (e.g., adenine). This adenine will fix at any point on the slide. However, each oligonucleotide has its own sequence. Thus, in order to fix the adenine only in the places where oligonucleotides beginning with an A must appear, a mask is placed to protect the places where adenine must not be present. At each step, the slides are soaked successively in solutions each containing a nitrogenous base of DNA, with a set of masks to synthesize oligonucleotides of the desired sequences at the desired places. For oligonucleotides of a length of 8 nucleotides, for example, 16 cycles (8 h) of synthesis are required, or 65,536 different possibilities of oligonucleotides. Each type of oligonucleotide is synthesized in 1 million to 10 million copies on a surface varying from 20 to 50 μm^2. One major problem with the use of these microchips is their inability to detect and correct the possible errors introduced during the *in situ* synthesis on the surface. Also, only shorter DNA molecules can be synthesized (generally 20 nucleotides).

Deposits of cDNAs

The cDNA probes that are to be deposited (on filter or glass slide) are amplified by PCR from a bacterial or plasmid library and then organized on microtitration plates of 96, 384, or 1586 wells. Standards for quality control of the hybridization can also be integrated with this collection of cDNA probes.

The cDNAs are then transferred on to the matrix constituting the support. Most often, the concentration of cDNA probes is standardized before the deposit in order to limit the deviations of DNA quantity from one well to

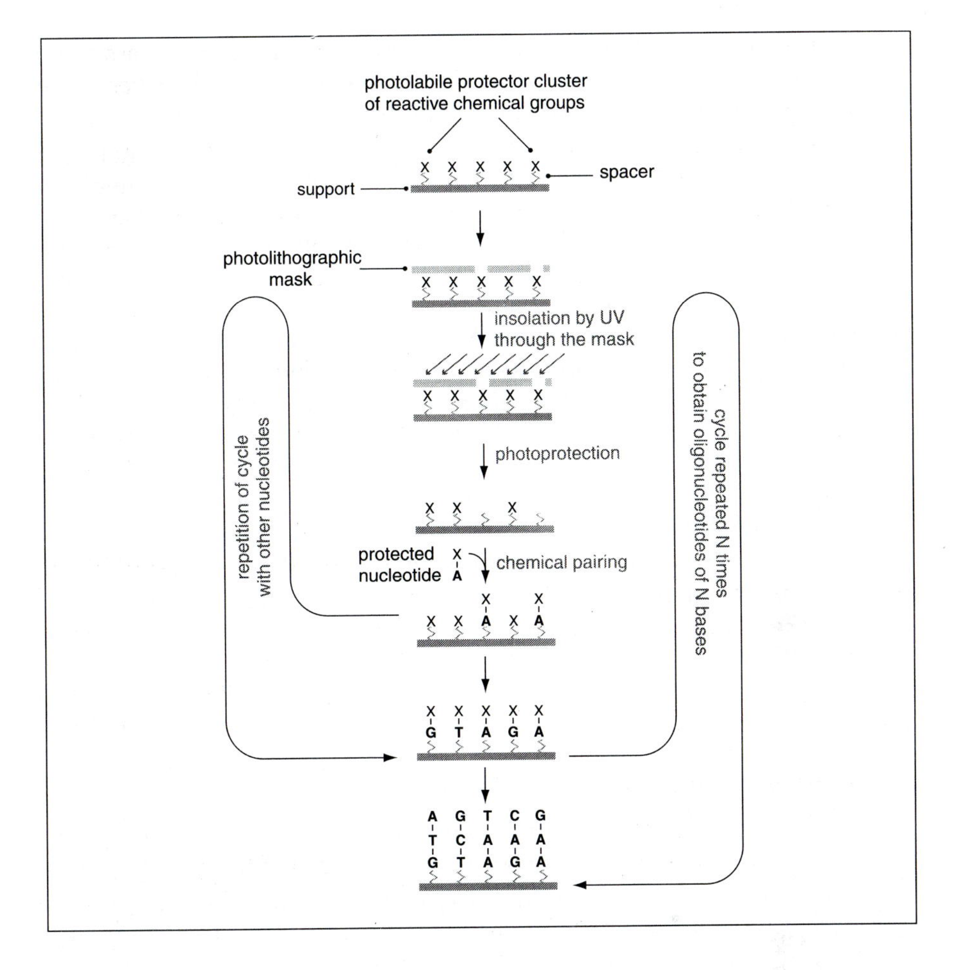

another. The quantities deposited depend on the size of the surface and range from one nanogram in the case of manual deposit to some picograms in the case of automated deposits. Glass slides need to be coated beforehand to allow the fixing of DNA molecules. The principal types of coatings used at present are epoxy, poly-L-lysine, silane, and aldehyde. In the case of poly-L-lysine or silane coatings, the cDNA is deposited and linked with a covalent bond to the support by an exposure to UV rays. In the case of an aldehyde coating, the primers used for the PCR are aminated, which allows the formation of covalent bonds with the support. After the cDNA is fixed, the slides are treated in various ways depending on the coating to allow saturation of the potential reactive sites on them. For example, in automated deposits a robot loads about 500 nl of purified PCR products extracted by means of a hollow needle from the wells and deposits 5 nl of each sample by touching the needle to each of the slides. Using a single sample, a robot can make up to 100 deposits. The deposits are generally 50-200 μm in diameter and are spaced 180-350 μm apart.

The cDNA filters are essentially nylon or nitrocellulose filters with a composition of quality identical to those used for hybridization of the Southern and northern type (see Profile 12). New types of supports made of plastic have been proposed, but their use is still limited. In manual systems deposits of a few mm^2 to 1 cm^2 are made for a total filter surface area of around 50 to 200 cm^2. The unit area of deposit of a cDNA, and thus the manual system of deposit, determine the final density of the cDNAs on the filter. In the case of automated deposits, a mechanical arm fitted with fine needles extracts and deposits cDNAs by capillary action. These deposits range from a picolitre to a nanolitre. The maximum density depends on the physical properties of the filter as well as on the length of radioelement emission that makes up the target DNA, which is used to reveal the presence of a hybridization (see later). This type of support allows for advanced miniaturization but does not equal the density of cDNAs on glass slides. Deposits on filters are followed by a buffer step or exposure to UV rays to fix the set of cDNA probes definitively and in a covalent manner to the support.

In order to be accessible after hybridization with target DNA, the cDNA probes must be denatured (rendered single stranded). NaOH at low concentration (0.1 to 0.5 N) is chiefly used to denature the probes after they are deposited.

Methods

Labelling and hybridization

The target DNA is prepared from RNA extracted from tissues, which can be compared to the contents of expressed genes. The RNA molecules are copied into single-stranded cDNA using a reverse transcriptase (see Profile 27). These cDNAs are then labelled radioactively (cDNA filters) or with fluorescence (glass slides), their synthesis occurring in the presence of a modified nucleotide base (see Profile 11). This step of reverse transcription requires quite a large quantity of RNAs (several microgrammes). In the case of biological material available in small quantities, this reverse transcription can be combined with amplifications of single-stranded cDNAs by PCR (RT-PCR). By this means a population of double-stranded cDNAs is obtained that can then be labelled by traditional means (see Profile 11).

With respect to hybridization of cDNA filters, the experiment conditions represent the protocols used for hybridizations of the Southern or northern type (see Profile 12), the filters being prehybridized in order to saturate the aspecific sites. The first series of filters is hybridized with the target DNAs of population 1 and the second series of filters (identical to the first) is hybridized with the target DNAs of population 2. For the hybridization, single-stranded or double-stranded target cDNAs (labelled radioactively) are put on a filter in a closed environment (tube, box, bag) in a hybridization buffer (see Profile 12). This step is followed by a series of rinses that are increasingly limiting to eliminate non-specific hybridization between the probe and target DNA.

One of the major differences between the use of cDNA filters and glass slides is that, in the latter, a single slide (bearing probe DNAs) is hybridized with a combination of labelled target cDNAs taken from two conditions that can be compared: the populations of target cDNAs of two situations are labelled each by different fluorochrome and the signal obtained will correspond to the fluorescence of targets of population 1 and targets of population 2. Each probe may hybridize with its target taken either from population 1 (labelled by a fluorophore) or from population 2 (labelled by another fluorophore). If the target is more abundant in population 1, it will have greater chance of hybridizing on the corresponding probe, which will then be charged with the fluorophore of target 1. The fluorescent targets are concentrated, placed in a hybridization solution, and mixed in equimolar concentrations before coming into contact with the slide. The slides are then covered with a coverslip and incubated for 16 to 24 h at a hybridization temperature ranging from 42 to 65°C. The targets not linked to probes are eliminated by rinsing. The slides are then dried by centrifuge.

In all the cases, a series of controls are necessary in order to verify that non-specific hybridizations have been eliminated.

Acquisition of data

The signals obtained on glass slides are emissions of fluorescence at the points of hybridization. The slides are first scanned in imagers, which detect separately each wavelength emitted specifically by each fluorophore. Colour images are obtained by this means. The scanners with glass slides are microscopes linked to a double or multiple laser emission system that excites the fluorophores linked to each target and receives the fluorescence emitted. The intensity of the signal of each spot for the two wavelengths of emission of the two fluorophores is then measured by quantification software, which can be used to determine for each probe DNA the ratio between its mRNA present in population 1 and that present in population 2.

In the case of cDNA filters, the reading of radioactive signals can be obtained by simple contact of filters on autoradiographic film. However, such film is often not sensitive enough to detect low intensity signals. In that case, scanners are used to read photosensitive screens on which the filters are placed in a cassette in the dark. The emissions of the radioelement used to label the target DNA will leave a footprint by transfer of energy on to the screen. These footprints are revealed by means of an imager that provides an image of hybridization signals. Each pair of filters is scanned separately. As with the glass slides, quantification software is used to assess the number of pixels associated with each hybridization signal, so that for each probe DNA the ratio of hybridization signals between the first filter (population 1) and the second (population 2) can be determined.

Data analysis

- Data processing

The raw data obtained after quantification of hybridization signals are often corrected before a more refined analysis is undertaken. Various types of correction can be made. Most often, the error is compensated for and the set of levels of expression measured in each of the conditions is standardized. The standardization is designed to balance out the intensities of hybridization obtained between the two populations of target cDNAs, as well as to allow comparison of the results between the different repetitions of an experiment. The standardization can be based on standards situated among the probe cDNAs or integration of the sum of all the signals measured in each of the conditions and relation of each value of expression to this sum of signals for each condition. In the case of glass slides, there may be a bias due to the fluorophores: this bias is clearly seen in an experiment with two identical samples of RNA labelled with two different fluorophores and then hybridized on the same slide. In this situation, it is rare to have equal intensity or ratios of 1. The biases have several origins, such as different physical properties of the fluorophores, unequal efficiencies of incorporation, or experimental variability.

The data thus corrected can be used directly in analysis of profiles of expression. They are also very often reduced by various modes of transformation (e.g., logarithmic or rank) before analysis.

- Reproducibility

Technologies based on cDNA filters or glass slides to quantify the level of expression of genes are highly effective. Nevertheless, they often vary significantly between several repetitions of the experiment. In fact, various steps of the technique may induce errors in the assessment of the level of transcripts: the deposit of probes (cDNAs or oligonucleotides), the steps of reverse transcription, PCR, and labelling of target cDNAs, the hybridization step, and the rinses. It is therefore necessary to apply methods of statistical analysis to this kind of data, for example analysis of variance (ANOVA), principal components analysis (PCA), or Student's *t* test.

- Methods of analysis

The simplest method is to compare the levels of transcripts between an experimental condition and an untreated condition called the reference. The corrected numerical values obtained for each gene can be placed on a scatter plot. This mode of representation gives for each gene the level of expression in each condition (condition A on one axis and condition B on the other), and also determines the level of regulation between the two states. The genes for which the points are placed along the bisector of the graph have identical levels of expression in each condition (x = y), while those that are far from this line are regulated. This method of analysis is very practical in identifying the most strongly regulated genes in a given state. Nevertheless, when the number of experiments increases (e.g., a kinetics of expression), such a graph is not suitable and other ways must be found to represent the data.

When several conditions are compared, a very large mass of data is soon obtained (several thousands of genes, several conditions, several repetitions). These data must be screened before they can be interpreted. The first screen involves statistical analysis to eliminate genes that do not show regulation under the test conditions. The regulated genes (which have "passed" the statistical test) are screened by various methods so that they can be grouped according to their profiles of expression. Different algorithms are available for this screening, but they all rely on the same principle: an analysis of the distance between the profiles of expression of genes. The genes are assimilated into vectors whose coordinates correspond to the levels of expression or ratios of expression obtained in different experiments. These "vector-genes" are replaced in a space that has as many dimensions as the experiments taken into account. The distance analyses group genes into pairs as a function of the smallest distances measured in the different dimensions (hierarchical clustering) or define several criteria that group similar genes (k-means, self-organizing maps). The groups can then be hierarchized on clades placed on a consensus tree that takes into account the distances between the different profiles of expression. These trees are most often associated with a colour matrix that indicates the level of expression or level of regulation of each gene. In the case of matrixes of regulation, the colour code attributes a given colour to stimulations of genic expression (ratio of expression > 1) and a second colour to regressions of expression (ratio of expression < 1). The genes with unvarying expression are most often represented in black.

cDNA filters or glass slides?

cDNA filters and glass slides have their own advantages and disadvantages.

One of the advantages of cDNA filters is financial, whether in implementation or the acquisition of material needed for the completion of the procedure. Many laboratories already have work stations that allow hybridizations of the Southern or northern type as well as manipulation of radioelements. The modification of such a station to the use of cDNA filters requires some refining of the hybridization techniques (optimization of hybridization buffers and rinses, choice of the radioelement ^{33}P for high-density cDNA filters) and the acquisition of an apparatus that allows high resolution scanning. With respect to the preparation of filters, easier systems of manual deposit are available and many companies offer among their services the deposit of amplified cDNA using robots or ready-to-use filters for some model organisms (yeast, *Arabidopsis*, human). The ability to manually deposit up to several hundreds of cDNAs on filters makes it possible to test genic expression routinely for certain types of experiments. Another advantage of cDNA filters is that they can be used several times for various experiments. The target nucleic acids that have been hybridized with probe cDNAs can be dehybridized. These filters can thus be used 2 to 6 times, depending on their composition.

However, glass has advantages over nylon. It is a durable material that tolerates high temperatures and successive rinses. It is non-porous and the volume of hybridization can thus be minimal, which increases the kinetics of hybridization between targets and probes. Glass does not contribute significantly to error because of its low fluorescence. Finally, it allows hybridization on the network of two or several probes labelled with different fluorophores for serial or parallel analysis. Thus, comparative analysis of a target situation with a reference situation can be realized directly on a single glass slide.

Applications

Profiles of gene expression

The main application of this technique is the study of gene expression. In this way, profiles of expression or matrixes of regulation are generated when different experimental conditions are compared with a biological reference material. The technique is extremely effective for diagnosis and allows the researcher to follow, for example, the effect of biotic or abiotic stress on a given biological material. Similarly, kinetics of expression can be studied.

Detection of regulons

The clustering of genes according to their profile of expression suggests that the clustered genes present a coordinated expression. They respond in the same manner to different biological conditions tested and constitute what could be called regulons (group of genes subjected to the action of a single transcriptional factor). That supposes the presence of common regulatory sequences near these genes (promoters).

These clusters in regulons also make it possible to predict the function or potential cellular category of a gene with an unknown coded function if that gene is clustered with a set of genes with a known function (e.g., a single metabolic pathway).

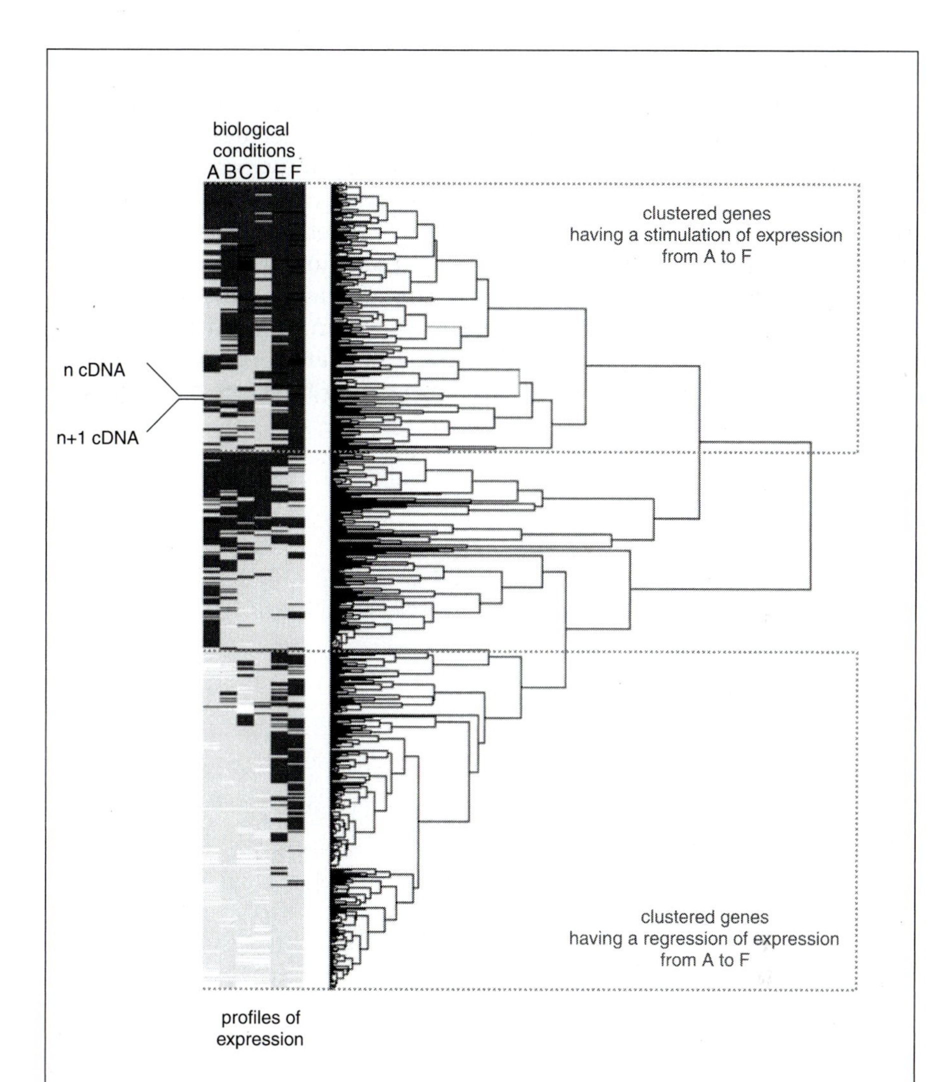

Graph of a clustering of profiles of expression of genes by hierarchical screening. The result of a hierarchical clustering analysis has two parts: a tree (at the right of the figure) that groups the genes step by step as a function of the similarity of their profiles of expression and a colour matrix (here shown in shades of grey, at the left of the figure) that describes these profiles of expression. On this graph, each line corresponds to a gene (=1 cDNA present on the filter or the glass slide) and each column corresponds to a level of expression of the gene in a given condition (A, B, C, D, E, or F). More exactly, there is a ratio of level of expression of this gene between the test condition and the reference conditon. When the expression of the gene is stimulated (ratio >1), a black colour is attributed, and when the expression of the gene is regressed (ratio <1), a whitew colour is attributed. The colourcode for the analysis is always represented in an inset. The tree on the right is the result of an analysis of distance between the profiles of expression of these genes, which allows classification of the genes according to their profiles of expression and clustering of similar genes.

Other applications

DNA microarrays may have uses in the analysis of genomes in correlation with the discovery of new medicinal or insecticidal molecules. For example, arrays of oligonucleotides can be used for the search for polymorphism or mutations in a whole genome, especially single nucleotide polymorphism (see Profile 56). DNA microchips can be used at different levels in the search for new active molecules, allowing the exploration of given metabolic pathways by identification of a series of co-regulated genes. When a new molecule needs to be tested, DNA microchips can be used to study the profiles of gene expression under the effect of that molecule in order to identify possible undesired effects. This expanding field has recently been named "pharmacogenomics".

---

[1]The terminology of these techniques is not yet completely established. The term *macro-* or *microarrays of DNA chips* can be considered generic. As there are two major processes of microchip production, the following can be distinguished: (a) macroarrays on filters of 1 dm^2 and more containing less than 100 spots/cm^2) and microarrays (on glass slides, of a few cm^2 and containing more than 500 spots/cm^2) with direct deposit of cDNA molecules on the support; (b) microchips with oligonucleotides with *in situ* synthesis of oligonucleotide probes on the support.

[2]The terms *probe* and *target* are used very precisely in experiments of molecular hybridization. The targets correspond to molecules of nucleic acids (here mRNAs) that are to be detected. These targets are found in a complex combination. The probes are molecules of nucleic acids that will serve to trap the target molecules by hybridization. The probes are purified molecules. In the case of DNA microarrays, the probes are fixed to the support and are not labelled, while the targets are in solution and are labelled. In the case of northern hybridization, the targets are fixed on nylon membranes and are not labelled, while the probe is in solution and labelled. DNA microarrays can be compared to an "inverse northern blot".

Karine Hugo and Sébastien Duplessis

V

Characterization of a Gene

Profile 23

DNA SEQUENCING

The refined characterization of a gene is done through sequencing, that is, through knowledge of the number, nature, and order of the nucleotides that make up the gene. Sequencing reveals, for example, the location of different restriction sites of the gene so that it can be better manipulated. Also, computerized translation of the nucleotide sequence into a sequence of amino acids can be used to confirm or propose a function of the protein coded by the gene.

There are several techniques of DNA sequencing, but here only the principle of enzymatic sequencing by incorporation of dideoxynucleotides is described. Dideoxynucleotides (ddNTPs) are modified deoxynucleotides that can integrate themselves into a DNA chain in synthesis but prevent the incorporation of the following nucleotide.

Principle

A nucleotide primer is hybridized with a fragment of single-stranded DNA in which the sequence is to be determined. From the 3'-OH end of the paired primer, a DNA polymerase synthesizes the complementary strand of the matrix DNA in the presence of deoxynucleotides. One of the deoxynucleotides must be labelled (radioactively or with fluorescence). The combination is then distributed in four tubes marked A, C, G, and T. Each of these tubes contains, in addition to the dNTPs, one of the ddNTPs. In each tube, the ddNTP added is incorporated into the fragments during their elongation.

Each time a ddNTP is incorporated in a new position, the elongation of the strand is stopped, which generates a set of molecules of different sizes that all terminate in the same ddNTP. If the dNTP and the corresponding ddNTP are added in adequate quantities, all the DNA fragments newly synthesized and terminating in this ddNTP will be represented. The contents of the four tubes A, C, G, and T are then analysed by acrylamide gel electrophoresis in denaturing conditions (or in capillary electrophoresis). The different newly synthesized, labelled fragments migrate as a function of their size and are separated at the nearest base. An autoradiography (in the case of radioactive nucleotides) or a fluorescence reading (in the case of fluorescent nucleotides) is done after migration of the molecules in the gel. The sequence 5'-P → 3'-OH of the newly synthesized strand can be read from bottom to top, comparing the relative positions of bands in the four tracks.

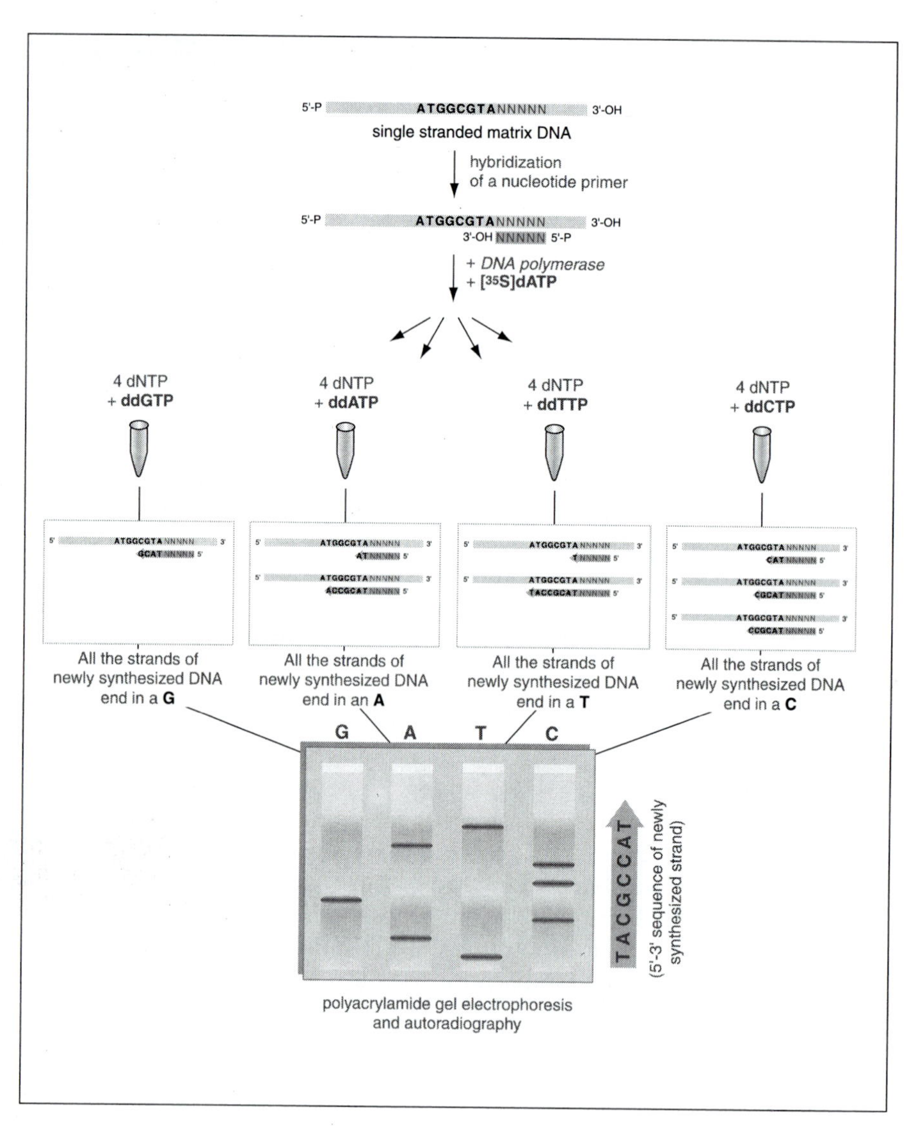

Automatic sequencing

During these analyses, the reading of the gel and acquisition of data are automatic. In this case, the molecules are labelled with fluorescent markers. During the migration of the gel, the samples are detected as they come out by a laser that identifies the labelled molecule. New systems based on capillary electrophoresis allow for better automation of the system. The samples are prepared manually and then deposited in an automatic loader. The apparatus automatically extracts the sample and loads it on the capillary. It

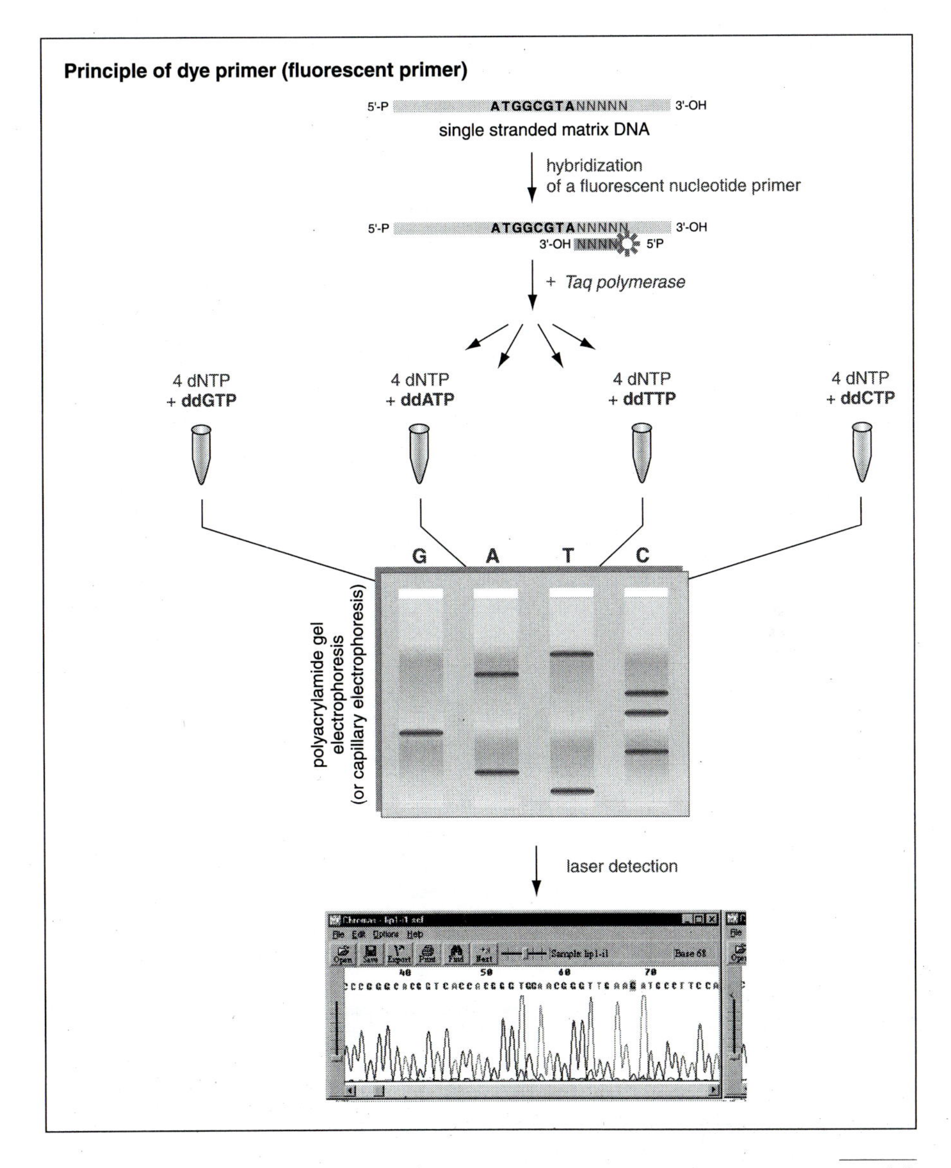

is no longer necessary to prepare acrylamide gels. Automatic sequencers can analyse up to 96 samples in a few hours. The sequencing reaction can also be automated; it is increasingly often carried out by robots attached to PCR machines.

In the dye primer method, the primer used in the sequence reaction is labelled by a fluorophore or dye. The reaction is carried out in four separate tubes, each containing the primer coupled with a fluorophore in the presence of a particular corresponding ddNTP. The DNA polymerase used

Principle of dye primer (fluorescent primer)

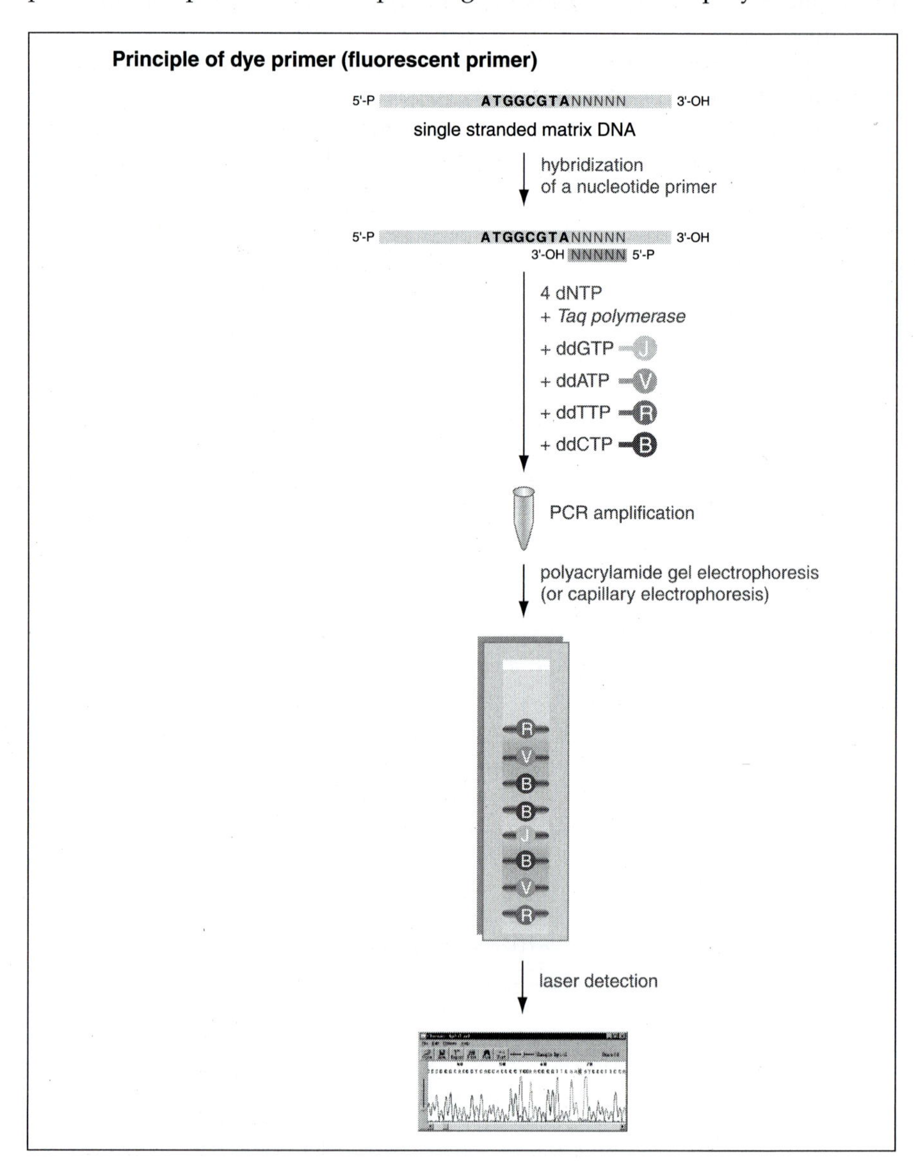

for the reaction is heat-stable and the polymerization occurs in a machine of the PCR type. with cycles comprising three steps: denaturation, hybridization, and extension (see Profile 24). At the end of the reaction, the denatured DNA fragments are separated on a polyacrylamide gel or a capillary electrophoresis matrix. The DNA fragments are automatically analysed when they are taken out of the gel by a laser system that allows the fluorophore to be identified. The results are transferred to a computer and a software program is used to get access to a chromatogram corresponding to the sequence

In the dye terminator method, it is the ddNTPs that are labelled, each with a specific fluorophore. All this is carried out according to the same principle as for the dye primer method, except that the reaction is carried out in a single tube. The advantage of the dye terminator method is that any primer can be used for the sequence reaction. It is thus possible to sequence a long fragment of cloned DNA by choosing the primer following from the end of the reaction of the preceding sequence. This is the method that is most commonly used at present.

Profile **24**

PCR: POLYMERASE CHAIN REACTION

PCR allows *in vitro* amplification of DNA sequences by repetition of elongation reactions in the presence of specific nucleotides and a DNA polymerase. The technique developed from the discovery of a thermophilous eubacterium living in the hot springs (70 to 75°C) of Yellowstone National Park, *Thermus aquaticus*, and the subsequent use of its polymerase, which is stable up to temperatures close to 100°C.

Principle

The principle of *in vitro* amplification lies in the repetition of three processes:

- denaturation of two DNA strands at high temperature (around 95°C) in order to obtain single-stranded DNA molecules;
- hybridization (annealing) of oligonucleotide primers complementary to a sequence of target single-stranded DNA (the temperature is then brought to a value between 40 and 65°C in order to allow for a better fixation of primers);
- the reaction of elongation (extension or synthesis) by a heat-stable DNA polymerase (*Taq* polymerase) from primers, realized at the optimal temperature of 72°C.

The products of this first cycle are then denatured by heat. The primers are again hybridized with DNA strands coming from the first amplification cycle, each strand serving as a template for the polymerase. At each cycle, the number of copies of the DNA fragment is doubled: 2^n molecules are thus obtained after n cycles, or for example 1,048,576 molecules after 20 cycles.

The PCR technique has literally revolutionized research in molecular biology and finds many applications in cloning and the study of gene expression as well as in research on genetic polymorphism.

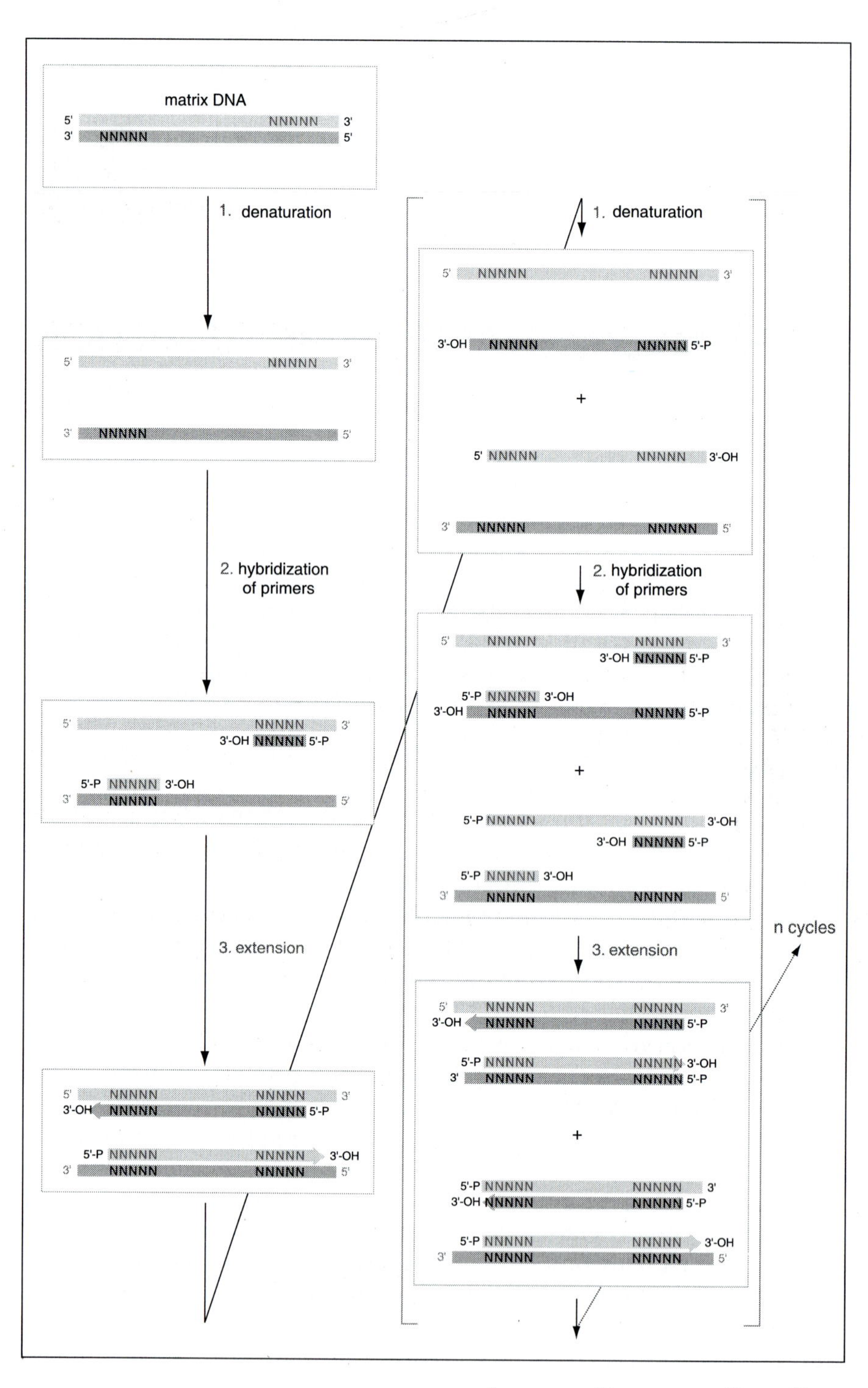

matrix DNA
1. denaturation
2. hybridization of primers
3. extension
n cycles

Profile 25

RACE: RAPID AMPLIFICATION OF cDNA ENDS

During the construction of a library of complementary DNA (see Profile 15), the 5'-P end of the sequence (the beginning of the cDNA) is sometimes absent because the reverse transcription (which begins the synthesis from the 3'-OH end) may not have been able to synthesize the cDNA completely (see Profile 27). Moreover, when cDNA is obtained by hexanucleotide random priming, one of the ends of the mRNA (5'-P or 3'-OH) is absent. Thus, the information present at the ends of the molecule is lost. One technique based on the use of PCR (see Profile 24) has been implemented in order to quickly complete the ends of the mRNA. This is the RACE technique or rapid amplification of cDNA ends.

The sequence of 5'-P or 3'-OH ends of an mRNA can be obtained according to the same basic principle: PCR-type amplification of the cDNA corresponding to the mRNA between a defined position on the molecule and one of the ends, 5'-P or 3'-OH. In both cases, it is necessary to know beforehand the partial sequence of the truncated cDNA studied.

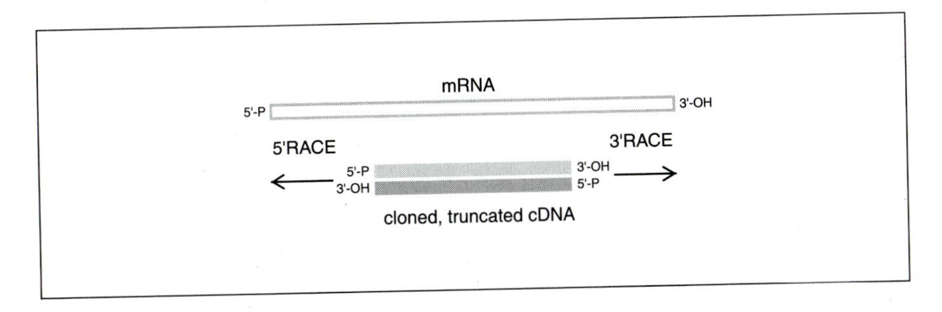

5′RACE

The principle consists first of synthesizing the cDNAs using reverse transcriptase, from a population of mRNAs extracted from a tissue known to accumulate the target transcript. The primer used for the functioning of the reverse transcriptase is an oligonucleotide that hybridizes specifically on the gene sequence to be cloned. After elimination of mRNAs (RNase H or NaOH treatment), the single-stranded cDNA obtained is modified by the fixation of an oligonucleotide of known sequence at the 3'-OH end. This sequence is often a poly-C or a poly-G. The single-stranded cDNA thus modified is then amplified by PCR (see Profile 24) using the specific primer of the gene and a complementary primer of poly-C or poly-G. The double-

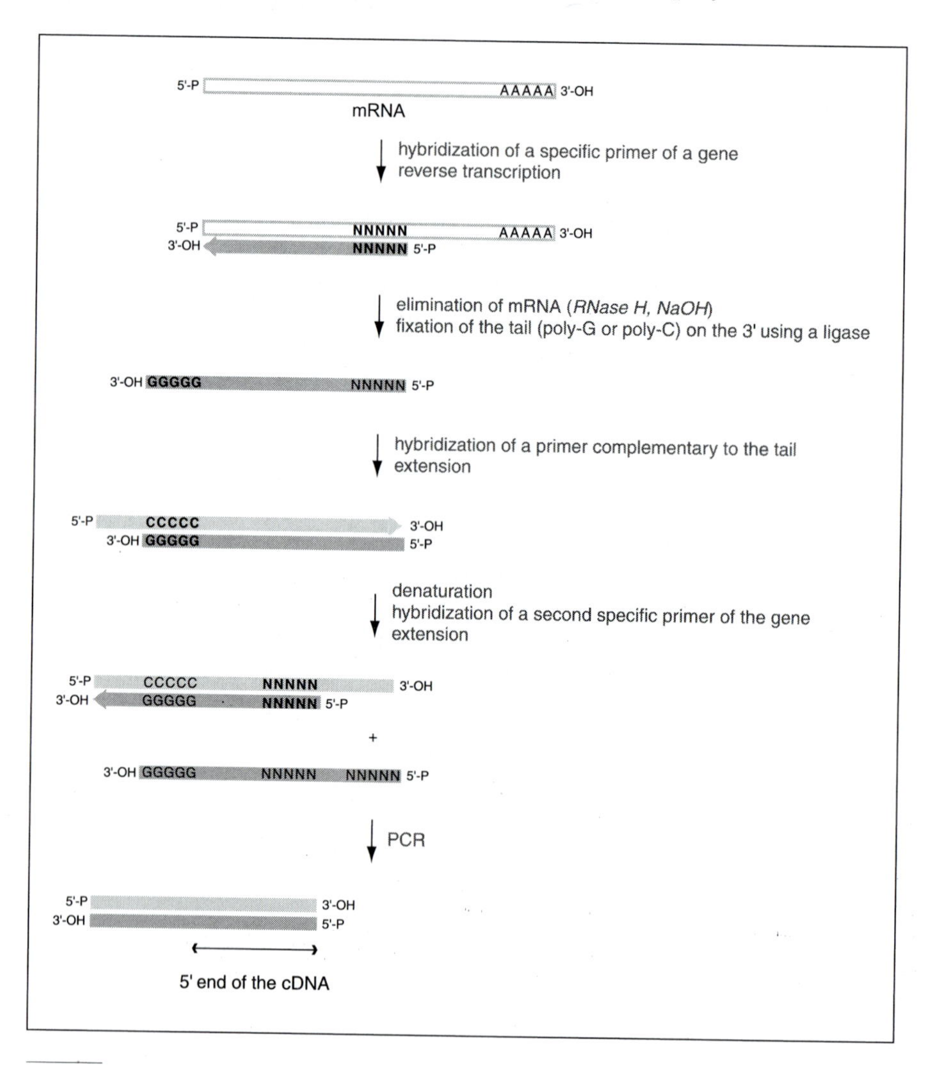

stranded DNA fragment obtained is then characterized by sequencing (see Profile 23).

3′RACE

The principle is as follows:

- Synthesize the strands of cDNAs by reverse transcriptase using as a primer a poly-T oligonucleotide containing at its 5'-P end a known supplementary sequence (adaptor oligonucleotide) and hybridizing on the poly-A end of the mRNAs.
- Amplify by PCR (see Profile 24) a cDNA fragment using a first primer hybridizing on the supplementary sequence of the adaptor oligonucleotide and a second primer hybridizing specifically on the gene. The double-stranded DNA fragment obtained is then characterized by sequencing.

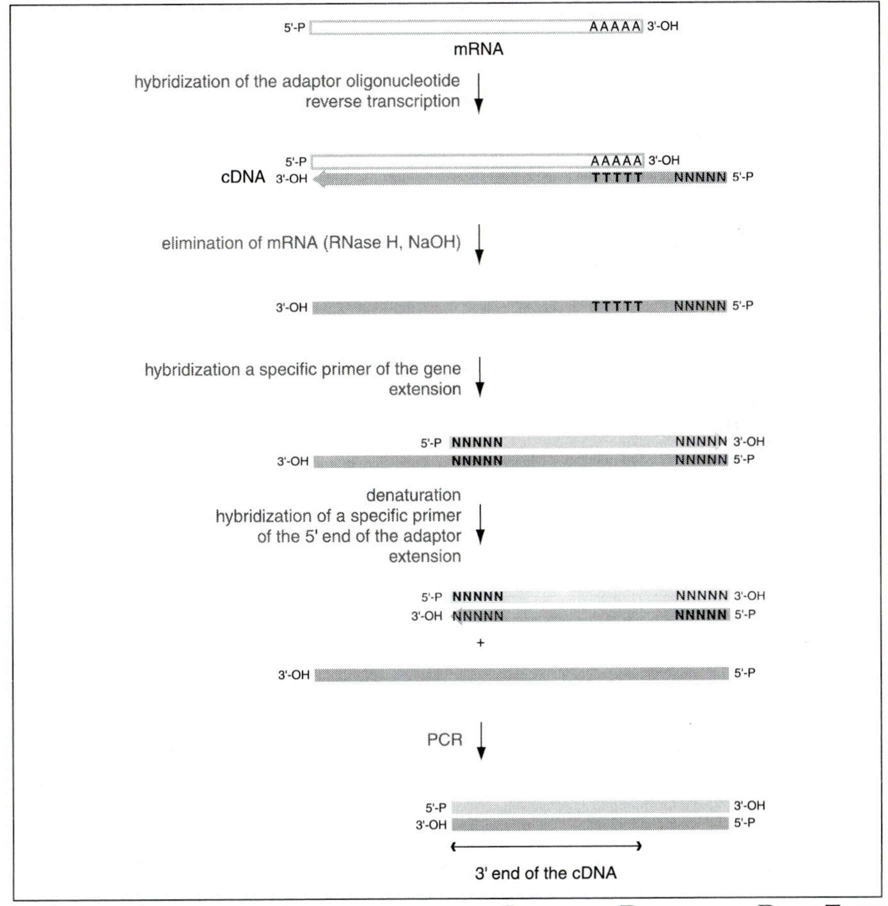

Profile **26**

GENOME WALKING BY PCR

Genome walking by PCR makes it possible to clone genomic sequences flanking a known DNA sequence. It is useful for example in the case of a gene tagged by a transposon (or, for plants, an insertion of T-DNA) (see Profile 45), or even for cloning the promoter of a gene for which the cDNA is not known. This technique can be used to characterize the sequence located upstream (5'-P) as well as the sequence located downstream (3'-OH) of the known sequence. It is a quick approach, highly sensitive, specific, and requiring little genomic DNA. It has the advantage of being operational for DNA of an individual (e.g., a mutant) for which a genomic library is not always available.

The genomic DNA is cleaved by a restriction enzyme leaving blunt ends. An adaptor presenting one strand longer than another is ligated at each end.[1] This DNA thus flanked is used as a template in a PCR reaction. The first primer is specific for the known DNA sequence. Depending on its orientation (sense or antisense), the extension will process upstream or downstream of this sequence (upstream in the figure). The second primer has the same sequence as the long strand of the adaptor. This primer will be able to hybridize only if one strand of DNA has previously been copied from the first specific primer of the known sequence. During the following cycles, there will thus be specific amplification of the sequence between the specific primer and the adaptor. The length of this sequence will depend on the position of the nearest restriction site (of the enzyme used to digest the DNA). Generally, four or five different enzymes are used in order to have PCR products of different sizes.

From this new sequence, a new primer can be designed and the process can be repeated in order to go further. Thus, the same population of genomic DNA fragments paired with adaptors can be reused, which constitutes a sort of genomic library for the genome walking.

[1]The 3'-OH of the short strand of the adaptor is blocked by an amino group in order to prevent a non-specific elongation of this strand with respect to the long strand during the PCR.

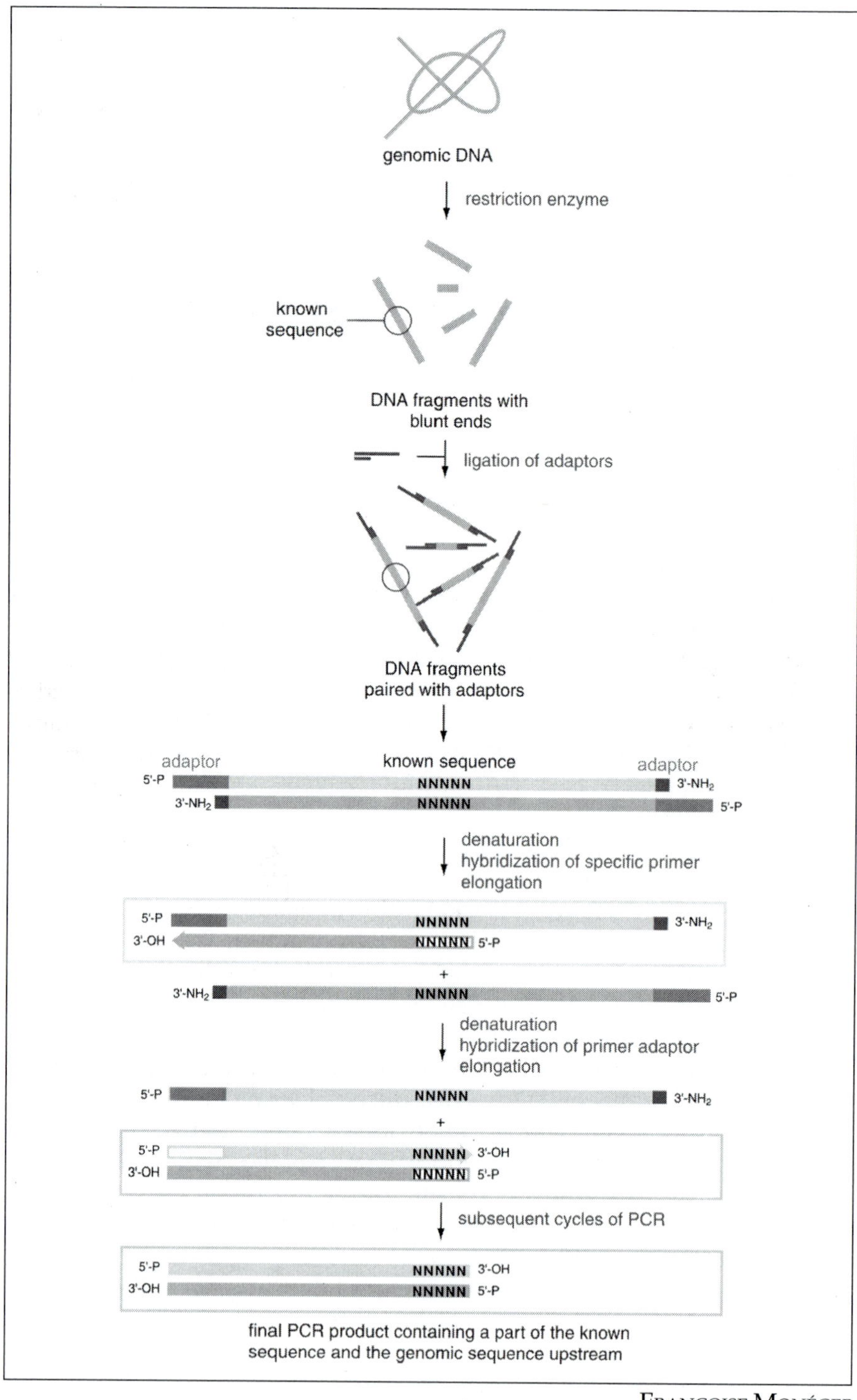

genomic DNA
restriction enzyme
known sequence
DNA fragments with blunt ends
ligation of adaptors
DNA fragments paired with adaptors
adaptor
known sequence
adaptor
5'-P
NNNNN
3'-NH2
3'-NH2
NNNNN
5'-P
denaturation
hybridization of specific primer
elongation
5'-P
NNNNN
3'-NH2
3'-OH
NNNNN
5'-P
+
3'-NH2
NNNNN
5'-P
denaturation
hybridization of primer adaptor
elongation
5'-P
NNNNN
3'-NH2
+
5'-P
NNNNN
3'-OH
3'-OH
NNNNN
5'-P
subsequent cycles of PCR
5'-P
NNNNN
3'-OH
3'-OH
NNNNN
5'-P
final PCR product containing a part of the known sequence and the genomic sequence upstream

Profile 27

RT-PCR: REVERSE TRANSCRIPTASE PCR

Conventional molecular hybridizations (Southern for DNA and northern for RNA, see Profile 12) are sometimes not sensitive enough to detect very small quantities of target DNAs or RNAs. This is the case especially for several mRNAs that are found in very small numbers in cells. PCR can be used to work with very small quantities of nucleic acids (see Profile 24). A technique called RT-PCR has been developed to detect and reveal the accumulation of a rare mRNA in an organ, tissue, or cell.

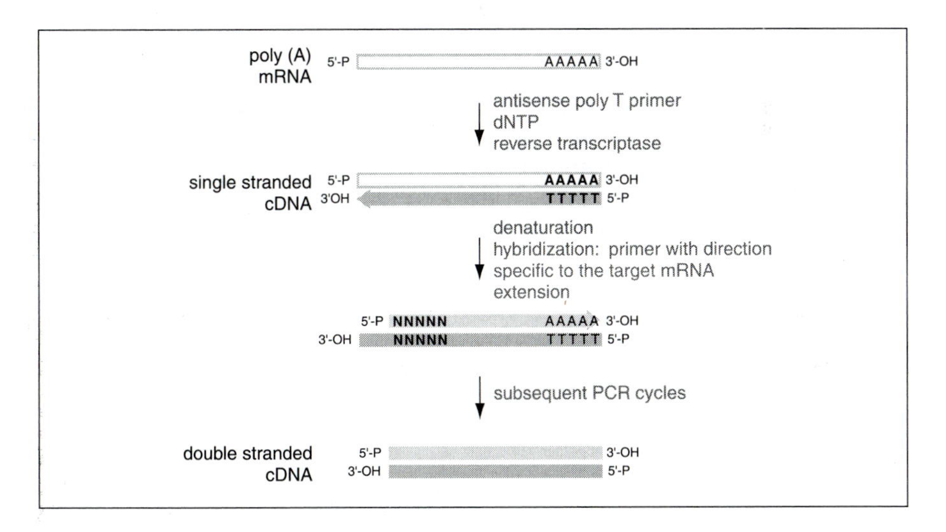

The principle consists of extracting the total RNAs of the tissues studied and copying them *in vitro* into single-stranded cDNAs, by means of reverse transcriptase. The DNA molecules obtained then serve as a template for a PCR reaction using a pair of specific primers of the target RNA sequence.

The PCR fragments obtained after the PCR cycles are visualized by gel electrophoresis[1] (see Profile 24). One of the difficulties of the technique is contamination of the RNA preparation by genomic DNA. In effect, the nucleotide primers are fixed on the single-stranded DNA from RNA as well as on the contaminant genomic DNA. One possible solution is to work from purified poly(A) RNAs. The contamination can also be prevented by treating the RNA samples with a deoxyribonuclease (very pure and lacking ribonuclease activity that could degrade the RNAs) in order to eliminate any trace of genomic DNA. On the other hand, where possible, a pair of primers that flank an intron sequence can be chosen: if the genomic DNA contaminates the preparation, the fragment of amplification obtained will be larger than the product of amplification obtained from the mRNA.

A modification of this basic technique can be used to evaluate the relative quantity of a given transcript in several samples and thus to evaluate the level of accumulation of that transcript following different treatments or in different tissues. The difficulty of interpretation arises from the fact that two things must be ensured: (1) that the same quantity of RNA is started with for each sample that has to be compared and (2) that it is a linear phase of amplification. Many internal standards are needed, such as co-amplification of the target gene and a control gene not regulated by the process studied; in this case, the amplification occurs in the presence of two pairs of primers, one recognizing the target gene and the other recognizing the control gene.

It is also possible to evaluate the quantity of RNA present in a given sample: this is quantitative RT-PCR. For this, a fluorescent marker must be used that allows the detection of PCR products as they are synthesized. The most elaborate and most specific technique uses as marker a fluorescent probe complementary to the target RNA. This probe of 20 to 30 nucleotides is paired covalently with a fluorophore at the 5'-P end and to a fluorescence inhibitor at the 3'-OH end. Moreover, it must be hybridized with the target RNA between the two primers used conventionally during a classic PCR. During the DNA polymerization phase, the probe is degraded (displaced by the DNA molecule being synthesized) and there is emission of fluorescence because the degradation leads to its fragmentation and thus separation between the fluorophore and the fluorescence inhibitor.

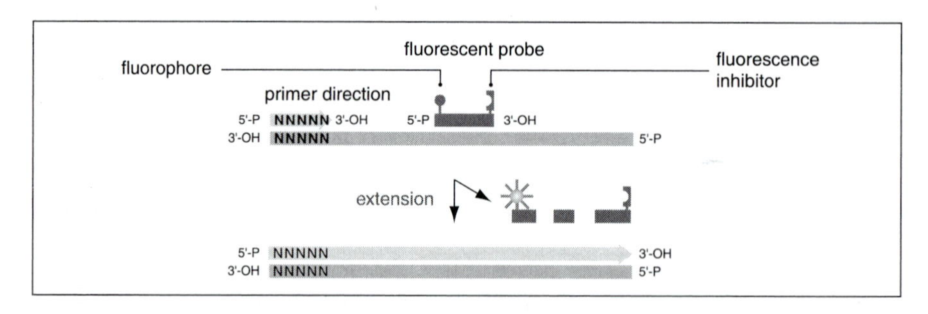

In the exponential phase of amplification, the emission of fluorescence is directly proportionate to the quantity of DNA synthesized during the reaction. The number of initial copies of the target RNA is thus determined by comparison with a standard curve generated from samples containing precise quantities of RNA studied, obtained by *in vitro* transcription. Some measures need to be taken to optimize the result. For example, a small RNA fragment (200 to 1200 nucleotides at most) must be amplified and a probe must be chosen that hybridizes stably with the target RNA under the conditions of the PCR. Moreover, it is essential to test two or even three dilutions of the sample to be analysed in order to get a result that falls within the limits of the standard range. Because of the specificity of this method, and by using different fluorophores, it is possible to follow the progress of several probes during a single PCR reaction. One of the applications of this technique is to use allele-specific probes to reveal genetic polymorphism.

---

[1]The DNA in the gel can also subsequently be transferred to a nylon membrane and subjected to molecular hybridization to ensure that the products of amplification obtained correspond to the sequence studied.

Profile 28

IN VITRO TRANSCRIPTION

The production of a given pure mRNA in large quantity is sometimes useful in studying for example the mechanism of RNA splicing or using RNA as hybridization probes.

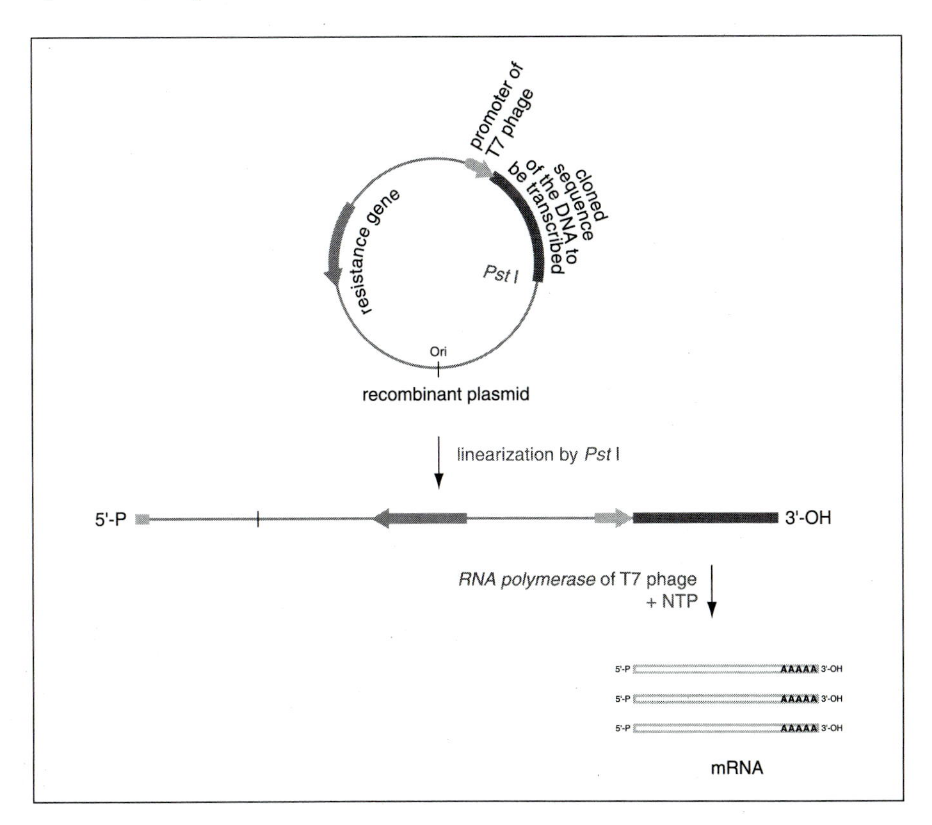

The sequence of DNA coding for the RNA to be studied is cloned in a plasmid or phagemid downstream of a promoter used by an RNA polymerase of the phage (e.g., T7, T3, or SP6). The DNA serving as template is first of all linearized at the end of the sequence to be transcribed by cleavage with a restriction enzyme. The RNA polymerase then initiates the transcription to the promoter and synthesizes the RNA strand complementary to the DNA sequence cloned in the presence of ribonucleotides. If one of the ribonucleotides added is labelled (radioactively or chemically), a labelled RNA is then produced. The *in vitro* transcription ceases when the RNA polymerase reaches the end of the linearized plasmid molecule.

Profile **29**

DETERMINATION OF TRANSCRIPTION INITIATION SITE

When the sequence of a gene is known and cloned in a vector, the site of initiation of the transcription of this gene can be determined by S1 nuclease mapping. To determine this site, the first base of RNA synthesized by RNA polymerase must be identified from the DNA template (see Profile 1).

The principle of this technique lies in hybridization in solution between mRNA of the gene studied (from a combination of mRNAs from the tissue studied) and a fragment of the genomic DNA of this gene, comprising the beginning of the coding part and a portion of the promoter sequence. This DNA sequence is first obtained by enzymatic restriction using restriction sites that flank the junction between promoter and coding sequence. In the example given, this is *Nco* I and *Spe* I. It is by analysing the sequence of this cloned DNA that the appropriate pair of enzymes is determined. This isolated DNA fragment is labelled radioactively at its 5'-P end (see Profile 11). The hybrid obtained occurs between the mRNA and the coding part of the gene. The promoter sequences remain single stranded. The use of a specific nuclease of single-stranded regions (S1 nuclease) allows the elimination of these sequences and just the double-stranded hybrid can be retained. The size of this hybrid is determined on a sequencing-type gel (see Profile 23) in order to identify its first nucleotide base, which represents the initiation site of the gene transcription studied.

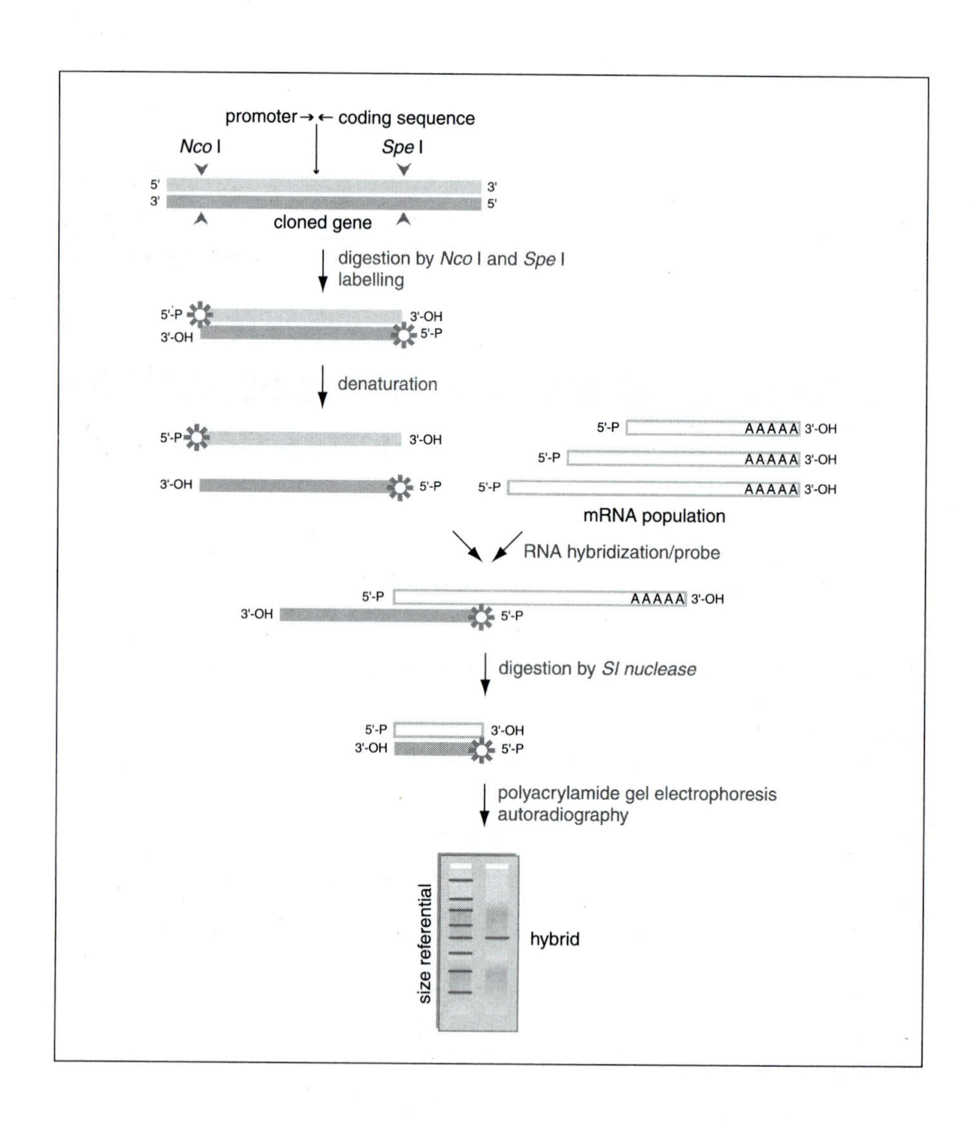
promoter → ← coding sequence
Nco I
Spe I
5'
3'
3'
5'
cloned gene
digestion by Nco I and Spe I
labelling
5'-P
3'-OH
3'-OH
5'-P
denaturation
5'-P
3'-OH
3'-OH
5'-P
5'-P
AAAAA 3'-OH
5'-P
AAAAA 3'-OH
5'-P
AAAAA 3'-OH
mRNA population
RNA hybridization/probe
5'-P
AAAAA 3'-OH
3'-OH
5'-P
digestion by SI nuclease
5'-P
3'-OH
3'-OH
5'-P
polyacrylamide gel electrophoresis
autoradiography
size referential
hybrid

Profile 30

FUNCTIONAL ANALYSIS OF PROMOTERS

The transcription of a gene is finely regulated in time and in space: a given mRNA will be synthesized only in certain cell types, under different biotic or abiotic stresses, and at a certain stage of development of the individual. The information needed for this regulation is carried by regulation proteins and by nucleotide sequences located upstream and downstream of the coding sequence (see Profile 1). It is the interaction between the regulation proteins and these DNA sequences that decides the transcription of a gene. The functional analysis of promoters allows the dissection of regions of DNA intervening in a spatio-temporal regulation of the expression of a gene.

Principle

The functional analysis of a promoter requires a knowledge of the nucleotide sequence of this regulatory region. A chimerical construction is realized, made up of the following: (1) a promoter subjected to functional analysis; (2) a reporter gene[1]; and (3) signals that allow the termination of the transcription. These chimeric genes are constructed *in vitro* in a plasmid by a series of cleavages by means of restriction enzymes, ligations, and cloning in *E. coli*.

The recombinant plasmid carrying the chimerical construction is used for the transformation of cells either in experiments of stable transformations and obtaining of transgenic organisms (see Profiles 34 to 38) or in experiments of transient expression (see Profile 38). The enzymatic activity of the protein coded by the reporter gene is then measured in the different transformed tissues or cells. The results are then interpreted as follows:

- If an enzymatic activity is measurable in a given tissue, that means the protein coded by the reporter gene has been synthesized, and thus the

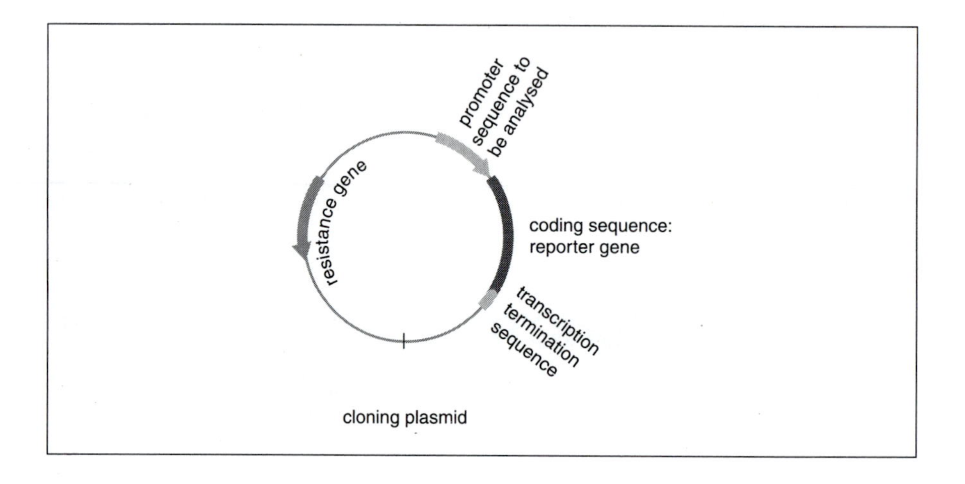

reporter gene has been transcribed under the control of the promoter part. This promoter is thus active in the conditions of analysis: it contains all the nucleotide sequences needed for the transcription of a gene in the tissue analysed.

- If no enzymatic activity is detected, the protein coded by the reporter gene has not been synthesized and the gene has not been transcribed. The promoter analysed is not functional in the transformed tissue. Either the nucleotide sequences for its proper functioning are lacking (sequence of the enhancer type) or certain nucleotide sequences present in this promoter prevent an active transcription (sequence of the silencer type).

By comparing the functioning of a promoter in different tissues, at different stages of development under different external constraints, the modalities of spatio-temporal expression of the promoter studied can be characterized.

Analysis by deletions

When a promoter directing the transcription of a reporter gene is functional after integration in a given tissue, the DNA sequences regulating the transcription can be delimited more precisely. To do this, the initial promoter is reduced in size and each new chimerical construction thus obtained is tested by transformation.

[1]For a definition of reporter gene, see Profile 38.

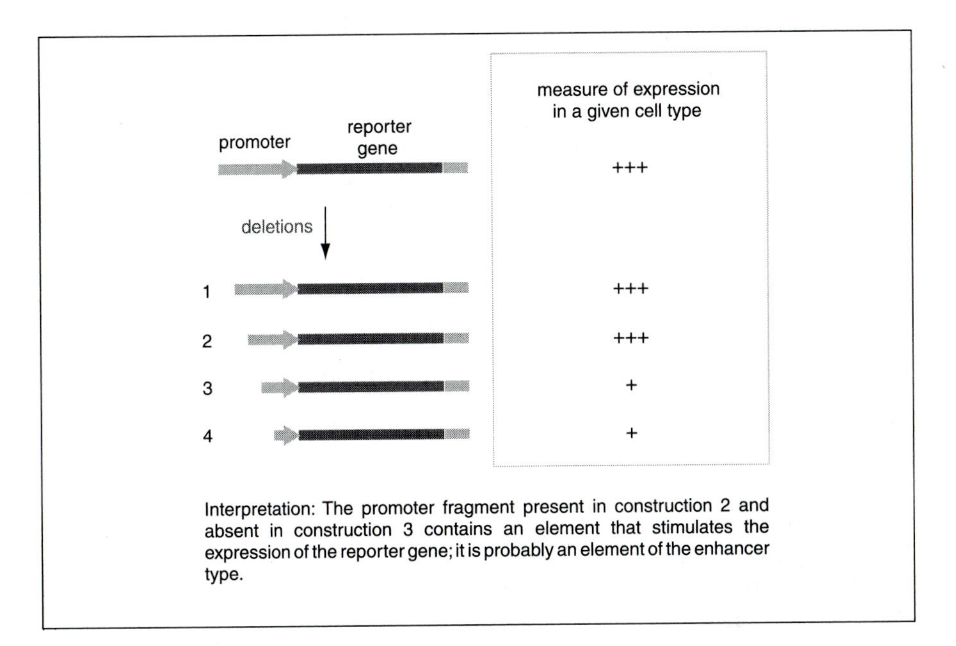

Interpretation: The promoter fragment present in construction 2 and absent in construction 3 contains an element that stimulates the expression of the reporter gene; it is probably an element of the enhancer type.

Profile 31

GEL RETARDATION

The expression of a gene depends on the fixation of regulation proteins (transcription factors, see Profile 1) on the promoter region. The genetic transformation technique associated with functional analysis of promoters can be used to delimit nucleotide regions of the promoter that may intervene in the regulation of gene expression (see Profile 30). The gel retardation technique makes it possible to detect proteins or proteic complexes that fix on these promoter regions.

The principle consists of the following: (1) extracting proteins from nuclei[1] of different tissues (generally a tissue expressing this gene and a reference tissue that does not express it); (2) hybridizing these proteins with the radioactive DNA fragment carrying the promoter sequence of interest; (3) depositing the hybridization mixture on non-denaturing gel to separate the DNA fragments by electrophoresis. If the fragment is hybridized with a nuclear protein, a complex is formed and its migration will be retarded in relation to a free fragment that has not been hybridized with proteins. The gel is autoradiographed and the electrophoretic mobility of DNA fragments that are complexed or not complexed is revealed.

This technique can be used to reveal the nucleoprotein complexes formed between nuclear proteins and a DNA fragment. It does not, however, allow the purification or characterization of proteins interacting with the promoter.

[1]Even when the regulation factors are nuclear, it is possible to use total proteins.

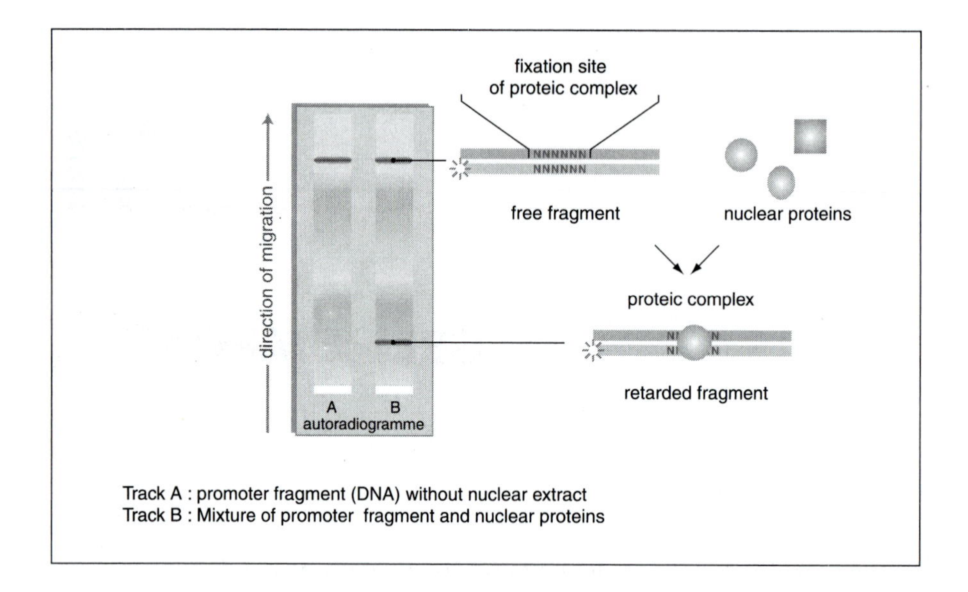

Track A : promoter fragment (DNA) without nuclear extract
Track B : Mixture of promoter fragment and nuclear proteins

Profile 32

DNase I FOOTPRINTING

Experiments on DNase I footprinting generally follow gel retardation experiments (Profile 31). They are used to define precisely the bases of the promoter that interact physically with a nuclear proteic factor.

The fragment containing the promoter sequence—and labelled radioactively at one of its ends—is hybridized with nuclear proteic extracts as before (see Profile 31). Then the combination is subjected to a moderated treatment with DNase I, which is an endonuclease that can cleave phosphodiester bonds within DNA chains of only one of the two strands. The result is a collection of fragments of different sizes, some of which are labelled at one end. The presence of a proteic complex conceals a possible site of restriction by DNase I and prevents the action of this enzyme. The products of digestion are then separated by electrophoresis on acrylamide gel with a high separating power (of the same type as sequence gels), then detected by autoradiography. When the bands obtained for a fragment of cleaved DNA are compared in the presence or absence of proteic factors, a region is revealed without a corresponding band at the site of fixation of the proteic factor, which by its linkage with DNA has protected the site of restriction with DNase I. In realizing a sequencing of the fragment studied and causing the migration of products of reaction in parallel with those obtained during an experiment in footprinting with DNase I, it will be possible to determine the sequence of the fixation site of the proteic factor.

Other techniques (not described here) can be used to confirm and refine the results obtained by DNase I footprinting (e.g., chemical footprints, protection by methylation). However, experiments that reveal the fixation of proteins *in vitro* do not necessarily reveal the situation *in vivo*. Within the cell, the concentrations of regulatory proteins and their DNA target are very different from those used *in vitro* and there is especially a great surplus of non-ligand DNA. Moreover, the DNA is thickly packaged in the form of chromatin, and many other proteins can interfere with the fixation of a particular protein on the target DNA sequence. Thus, techniques considered *in vivo* make it possible to analyse the fixation of regulatory proteins on portions of DNA, without greatly disturbing the integrity of the cell. The methods consist of modifying the DNA *in situ* by UV, chemically, or by enzymes (nucleases) under mild conditions. The DNA is then isolated and the modified nucleotides are detected on sequence gel.

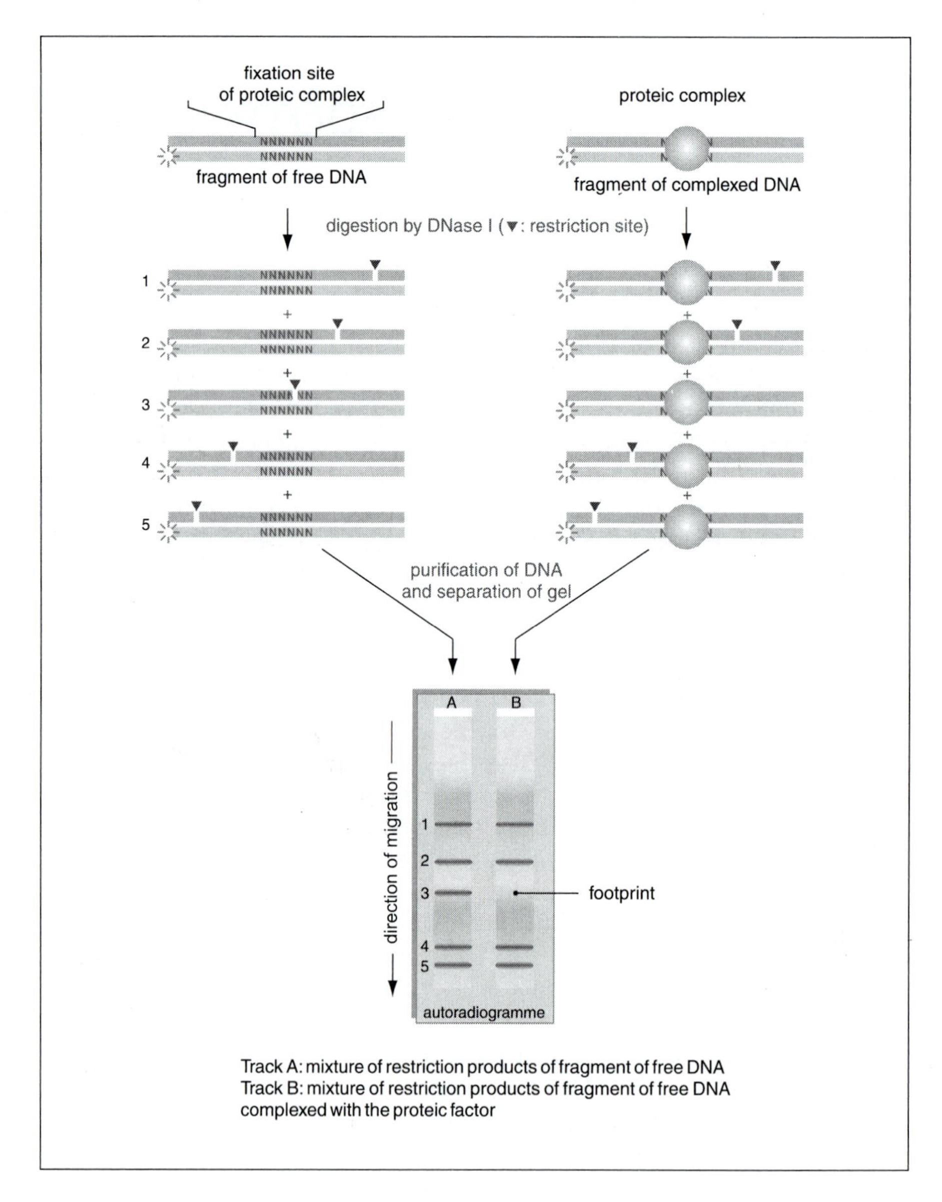

Track A: mixture of restriction products of fragment of free DNA
Track B: mixture of restriction products of fragment of free DNA complexed with the proteic factor

VI

Genetic Transformation of Eukaryotes

Profile 33

GENETIC TRANSFORMATION OF PLANTS BY *Agrobacterium tumefaciens*

Agrobacterium tumefaciens is a soil bacterium responsible for crown gall (the crown is the intersection of root and stem), a disease affecting dicotyledonous plants, which is manifested in uncontrolled division of cells at the site of infection and the development of a plant tumour. During the process of infection, complex molecular mechanisms are involved in the stable integration of a DNA fragment of bacterial origin into the plant genome. This natural mechanism of DNA transfer has been "domesticated" for biotechnological purposes and the co-culture of *A. tumefaciens* is today undoubtedly the most commonly used method of genetic transformation of plants. A few years ago, *A. tumefaciens* was also used for genetic transformation of yeasts and filamentous fungi, as well as human HeLa cells.

Description of the natural process

Agrobacterium tumefaciens has one large plasmid, the Ti (tumour-inducing) plasmid, which carries a significant part of the information needed to transfer bacterial DNA into the host plant genome.

The T region, present on the Ti plasmid, corresponds to the DNA sequence that will be transferred into the plant genome as T-DNA. This region is delimited by two direct repetitions of 25 base pairs, the borders. Between these borders are located the following: (1) genes inducing the development of the tumour (oncogenes), the expression of which leads to disequilibrium in the hormonal balance of the transformed cells, thus changing their growth characteristics (genes for biosynthesis of auxins and cytokinins, plant hormones that are responsible for the uncontrolled multiplication of plant cells) and (2) genes that direct the synthesis of opines,

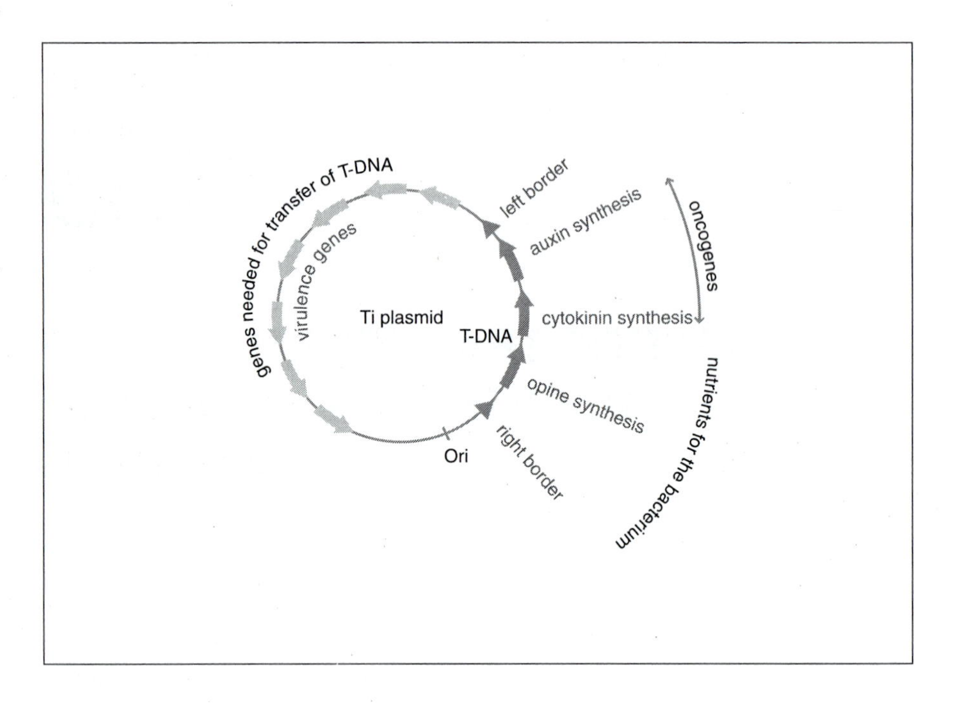

which are amino acids linked to a sugar, usable by the bacterium and not by the plant (e.g., octopine, agropine, nopaline). It has been shown that the DNA sequences located between the two borders are transferred into the genome of plant cells under the action of proteins coded by the genes of the *vir* region. The *vir* region carries a series of virulence genes that code for proteins intervening in the mechanisms of transfer of the T-DNA into the plant cell (in the single-stranded form), its integration in the genome of the host plant, and restoration of the single-stranded T-DNA of the plasmid by replication. The expression of virulence genes is triggered by growth factors, molecules liberated by cells of the plant during an injury. However, although our understanding of interaction between *A. tumefaciens* and the plant cell is rapidly improving, the molecular mechanisms involved in the transfer of T-DNA are not yet fully characterized.

Biotechnological uses of *A. tumefaciens*

The interaction between *A. tumefaciens* and the host plant constitutes a natural mechanism for transferring DNA into plants. This mechanism has been modified to become a versatile tool of genetic transformation. In fact, only the borders of the T-DNA are needed for its transfer into the plant genome. Because of this, the Ti plasmid could be disarmed by deletion of oncogenes present on the T-DNA of the wild strain: the new strain thus loses its capacity to induce tumours. The target genes as well as the sequences

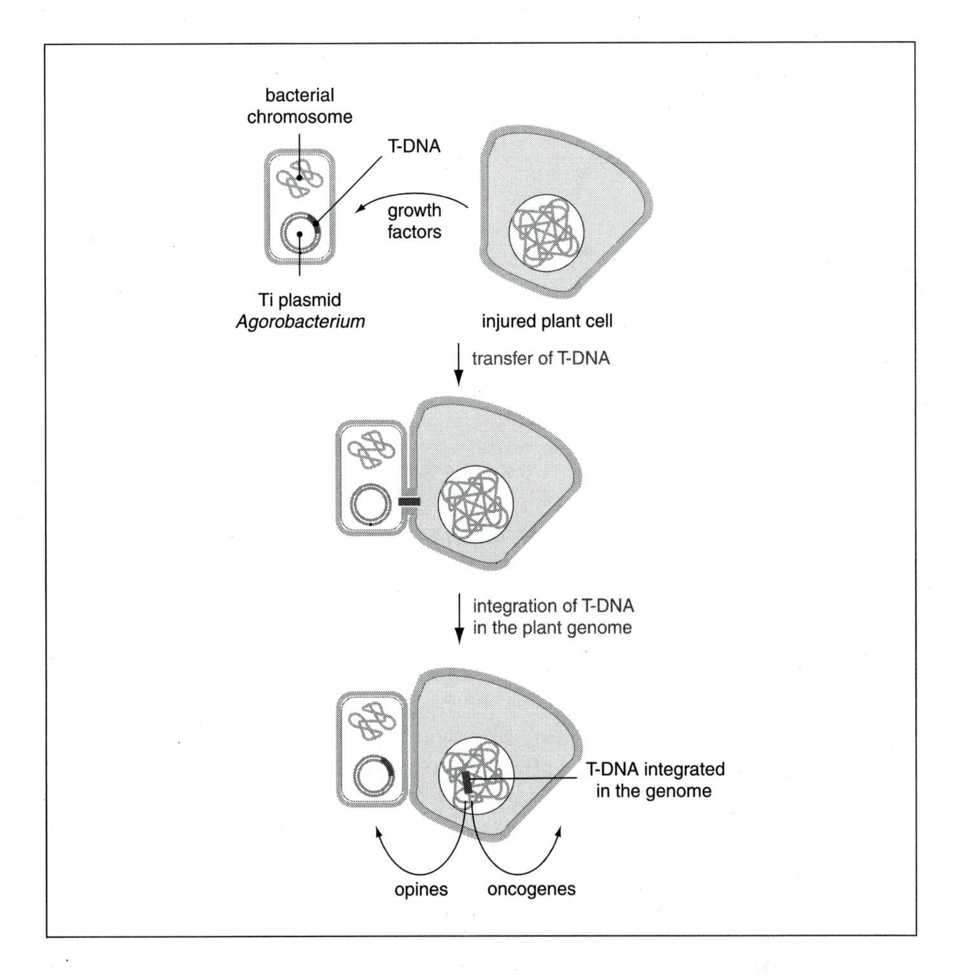

needed for the realization of the genetic transformation are located between the borders of the T-DNA. Most often, this modified T-DNA is placed on a second small plasmid called binary vector, while the *vir* functions are provided in *trans* by the disarmed Ti plasmid.

Conventionally, a binary vector carries an origin of functional replication in *A. tumefaciens* as well as *E. coli*, markers for selection for plants and bacteria, the border sequences of T-DNA, and a multiple cloning site. Between the right and left borders are generally found a selection gene that allows selection of genetically transformed plant cells as well as the chimerical construction that is sought to be introduced in the plant.

The target gene may be a gene already present in the plant genome, the function of which must be studied by varying its level of expression in the transformed plant and evaluating the effects on its phenotype. Depending on the type of chimerical construction (e.g., the orientation of the coding

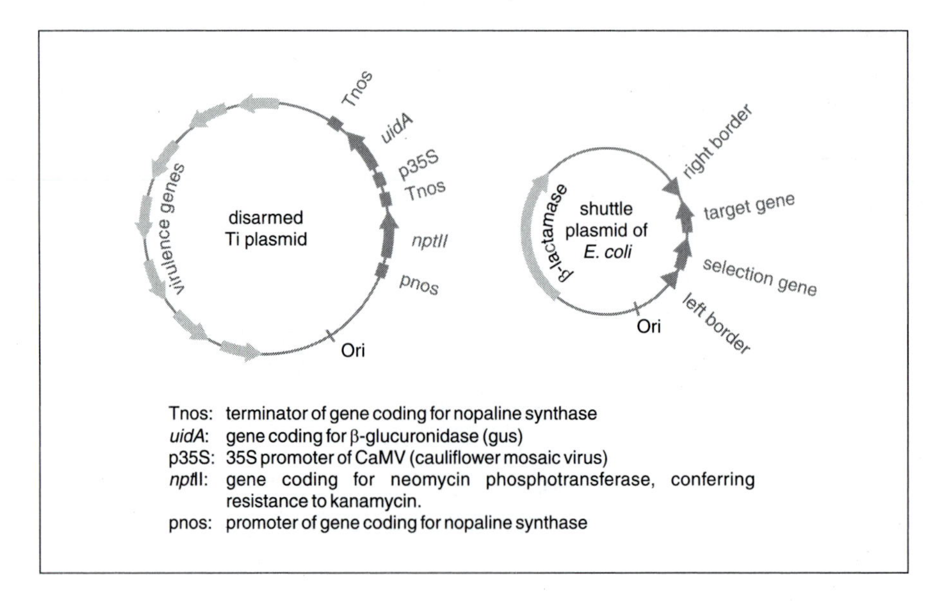

sequence, its head-to-tail duplication), the gene introduced could be expressed ectopically in the transgenic plant (at a step in its development or in a tissue in which it is not expressed in the normal plant), be overexpressed, or lead on the contrary to the underexpression of the corresponding endogenous gene by antisense effect, co-suppression, or RNA interference (RNAi, see Profile 46). The sequence could be of agronomic interest (conferring, for example, tolerance to insects). Or it could be a promoter that is useful in studying the specificity of expression in the development of the plant, this promoter being fused with a reporter gene (see Profile 38) in the chimeric construction. The resistance gene conventionally codes for a protein conferring resistance to an antibiotic. Today, new vectors have emerged that use other agents of selection (resistance to herbicides, resistance to heavy metals) or even mechanisms that allow the excision of a gene for resistance to an antibiotic after the selection phase (e.g., the recombinase system). This last strategy allows for the possibility of several successive transformations using a single selection agent.

Transformation

To transform plant cells, explants (leaf discs, stem internodes) are put in contact with the dilute culture of *A. tumefaciens* in controlled conditions. It is during this co-culture step that the modified T-DNA is transferred to the plant cell and integrated into its genome. The decontamination step consists of removing by successive rinses most of the henceforth useless agrobacteria. The explants are then maintained in a suitable medium containing the antibiotic that allows the selection of genetically transformed plant cells. The transformed cells are then multiplied and put into a medium in which they are regenerated into whole plants through adventitious budding or somatic embryogenesis.

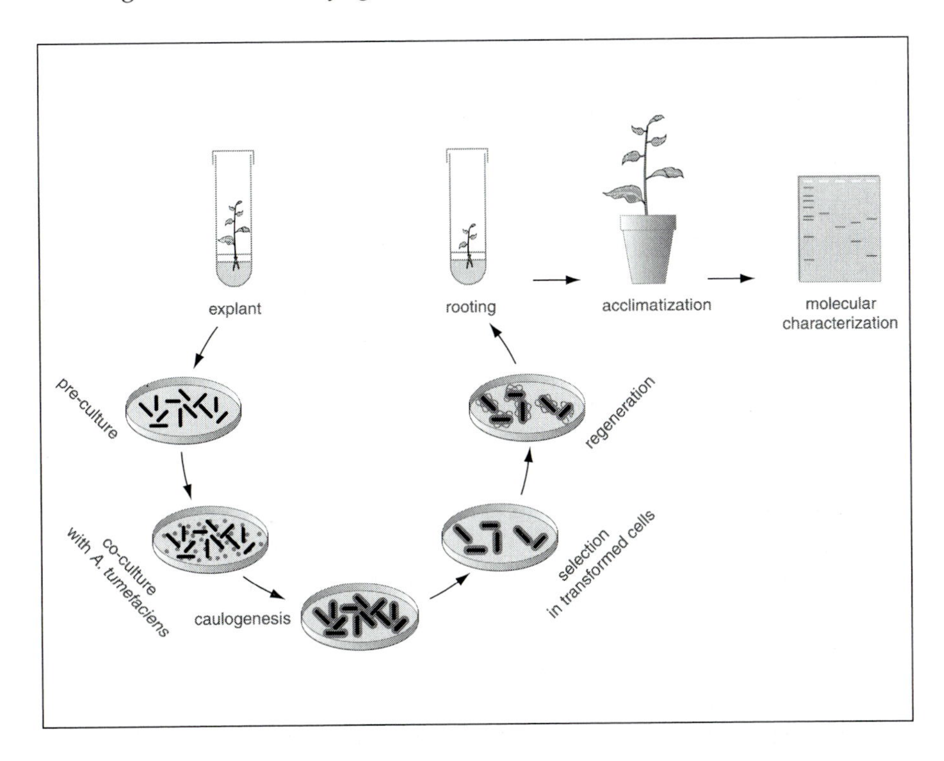

The transgenic plants are then characterized at the molecular level in order to verify the integration of the T-DNA in the genome and estimate the number of copies transferred (Southern hybridization) as well as to evaluate the profile of expression of the gene or genes introduced at the level of mRNAs (northern analysis) or proteins (western analysis), or their biological activity. The phenotype of transgenic plants is compared to that of unaltered control plants. Controlled crosses are realized in order to obtain in the progeny lines that are homozygous for the transgene. Multi-transgenic plants are obtained when plants that are transformed with different transgenes are crossed together.

A high-throughput genetic transformation technique has been developed in the model plant *Arabidopsis thaliana*. This method, which skips all the steps of *in vitro* culture required in the conventional protocols, consists of a brief soaking of plants with floral buds in a solution of the agrobacteria. The selection occurs ultimately at the level of germination of seeds produced by these plants. This method can be used conventionally to introduce a gene into a plant, but it is especially used as a method of random mutagenesis by tagging of genes with T-DNA: several tens of thousands of transgenic lines have already been produced. In using adapted binary vectors, this method has also made possible the tagging for gain of function mutants. For this, the T-DNA carries near the right border a strong promoter directed towards the exterior, leading potentially to the activation of the transcription of genes present around the insertion site of T-DNA in the genome. Similarly, using this method with a T-DNA carrying near its right border a reporter gene without promoter, the promoter sequences near the insertion site can be identified. Thus, genetic transformation by *A. tumefaciens* has been a high-performance tool in the approaches of functional plant genomics.

Profile 34

DIRECT GENE TRANSFER INTO PLANT PROTOPLASTS

The principle of direct transfer of genes consists of mixing plant cells with DNA in solution. The DNA is not necessarily a transformation vector and the transgenes can be DNA fragments, which are called "naked DNA". However, the cell wall in plants (and fungi) is an obstacle to the integration of DNA in cells. Therefore, it is necessary to prepare protoplasts by digestion of the pecto-cellulosic wall that will be mixed with the DNA in solution. The application of appropriate physicochemical conditions to this mixture causes a transitory and reversible permeabilization of the plasma membrane that allows DNA molecules to penetrate the protoplasts. These DNA molecules reach the nucleus—by a mechanism that is still unknown—and are integrated by recombination with chromosomes of the plant cell.

Method

The DNA mixture used can be made up of the following: (1) a plasmid carrying a gene whose expression confers resistance to a selection agent in order to select transformed cells and (2) a plasmid carrying the chimeric construction of interest. This is called co-transformation. The protoplasts thus treated are cultivated first on a non-selective medium to allow them to synthesize a new cell wall and initiate the first cell divisions, then on a medium containing the selection agent in order to eliminate non-transformed plant cells. These surviving cells are then subjected to processes of regeneration in order to obtain resistant plants. The resistant plants are analysed in a fashion analogous to that described for plants obtained through transformation by *A. tumefaciens* to demonstrate that they are transgenic.

Three major techniques are used to introduce the DNA into the protoplasts: (1) use of polyethylene glycol (PEG); (2) electroporation; and (3) biolistics (see Profile 35).

Polyethylene glycol is a polymer that causes fusions between plasmalemmas. It allows the physical connection of DNA molecules and plasmalemma, and local and reversible destabilizations of the plasmalemma. The result is the penetration of DNA molecules into the cytoplasm of protoplasts.

Electroporation consists of applying to the DNA-protoplast mixture an electric shock that also destabilizes the structure of the plasmalemma in a reversible manner. The procedure allows the DNA to be integrated into the cells. This protocol, under certain conditions, increases the efficiency of the transformation by a factor of 100 to 1000.

Profile 35

DIRECT GENE TRANSFER BY BIOLISTICS

Principle

Biolistics was developed in order to introduce genes into whole cells rather than protoplasts because of the difficulties of regeneration encountered in some plant species. It is also applicable to animal cells. The principle is to forcibly deliver DNA into tissues by propulsion. The propulsive force is achieved either by explosion of gunpowder in a bullet or by decompression of a gas (usually helium).

Method

The naked DNA to be introduced is first precipitated on tungsten or gold micro-particles of around 1 μm diameter. These micro-projectiles are propelled by means of a macro-projectile on to the cells. After the macro-projectile is arrested so that it will not destroy the cells, some of the micro-projectiles penetrate the cells, transporting with them the DNA. This DNA, as in the case of direct transfer of genes into protoplasts, must reach the nucleus, integrating itself in a part of the genome and expressing itself there. The explants, after bombardment, are cultured on a selective medium in order to regenerate transgenic plants. This approach makes it possible to obtain numerous transgenic plants for species that have so far been difficult, especially maize and wheat. Several types of equipment are available in the market for propulsion with bullets or using gas. Miniaturizations have been achieved that make it possible to use biolistics to transform animal tissues *in vivo* under anaesthesia.

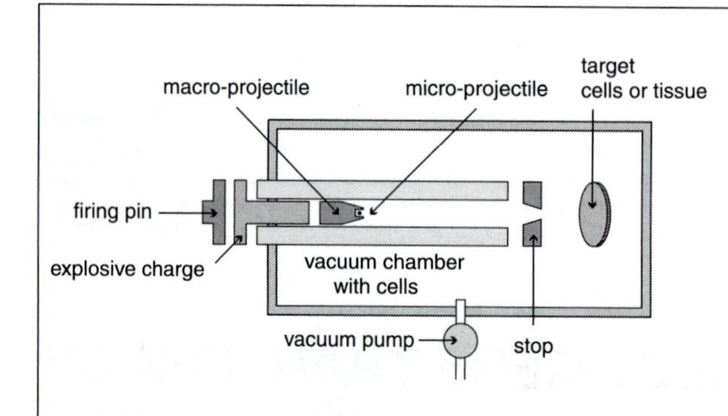

Drawing of a gun with particles by explosive charge (Klein et al., 1987, High-velocity micro-projectiles for delivering nucleic acids into living cells. Reproduced with the permission of Nature, 327: 70-73. Copyright 1987, Macmillan Magazines Ltd.

Profile 36

GENETIC TRANSFORMATION OF ANIMAL CELLS

The introduction of a DNA sequence in an animal cell makes it possible to analyse the role of that sequence *in vivo* by determining the function of a gene or identifying the regulatory sequences controlling the expression of the gene. The researcher must therefore choose the most appropriate cell line for the production of a given protein or for the study that is being carried out, as well as the most suitable vector for the cell line selected. The transfer of genes may allow their transient expression from non-integrated plasmids (see Profile 38). It is also possible to select cells in which the gene transferred is integrated. Some vectors can be maintained at the autonomous episome stage (free DNA molecules in the cytoplasm). This is the case of vectors derived from virus SV40 in COS cells and those derived from the Epstein Barr virus in primate cells.

All the vectors used must have the following essential elements:

- a bacterial origin of replication and a marker of resistance to an antibiotic that allows the amplification of the plasmid in bacteria;
- eukaryotic sequences that control the initiation of the transcription (promoters) and the maturation of transcripts (signals of splicing, polyadenylation...) derived from cellular or viral genes;
- a selection marker that allows the survival of recombinant eukaryotic cells if they are cultured in a selective medium;
- the gene to be expressed in the form of a complementary DNA or even a genomic DNA.

Two experimental approaches have been developed to introduce DNA into eukaryotic cells: one is the transfection of a vector of expression partly in the form of naked DNA and the other is a process of viral infection. In the latter case, the technique of infection depends on the type of virus and host cells. The principal viral vectors are derived from adenoviruses and retroviruses, as well as others such as Semliki Forest Virus.

The techniques used

Several methods of transfection of animal cells by naked DNA are presented in this section. The efficiency of penetration and expression of the foreign DNA may vary from 1 to 100 (or more) depending on the type of cells used, the nature, the quantity, and the state of the recombinant DNA and various physicochemical parameters of the introduction of the DNA. The choice among these techniques depends mostly on the nature of cells used (e.g., the method using calcium phosphate is ineffective for lymphocytes) and on the type of expression desired (a high rate of stable expression is obtained by electroporation). Very often, the DNA to be introduced is associated with positively charged molecules (calcium, cationic polymers) so as to favour its contact with the plasmalemma negatively charged on its surface.

Use of calcium phosphate

A mixture of DNA, phosphate buffer, and a calcium salt leads to the formation of calcium phosphate–DNA precipitation. The precipitate spontaneously enters the cells but this step is greatly improved by an osmolytic shock applied to cells by means of a glycerol or dimethylsulfoxide (DMSO). This method is much more rarely used since the use of polycationic polymers.

Use of cationic polymers

Hybrid molecules comprising a lipid region and a cationic region offer the advantages of both types of chemical properties. Many formulas are commercially available. Cationic lipids are easy to use. Since their efficiency varies with the cell lines, the choice must be based most often on tests or on published results. Certain cationic lipids are costly and have a limited stability.

- DEAE (diethyl amino ethyl) dextrane is a polycation that facilitates the adsorption of DNA on the plasmalemma and favours its entry into the cells. Moreover, it protects the DNA against the effect of nucleases and thus increases its rate of expression in the cell. It is also used to increase the efficiency of viral infections. The cells contain a large number of vectors in their nuclei, but few copies are integrated in the genome of the cell. However, DEAE dextrane can be toxic to cells. The optimization of transfection conditions depends on the sensitivity of host cells to this polymer. Nevertheless, the efficiency of transfection is high, since up to 25% of transfected cells can be obtained.
- Polyethyleneimine (PEI) is a polycation that associates easily with DNA: the complex thus penetrates cells spontaneously. The efficiency of PEI depends largely on its size. It is stable, economical, reliable, and

effective in a large number of cell types. It is easy to use. Exgen 500 is an optimized form of PEI. Other polycations can also be used, but they are not described here.

Use of liposomes

The technique consists of trapping the foreign DNA in artificial membranous lipid vesicles. The vesicles penetrate the cell by endocytosis. The infectious power of the virus with DNA such as SV 40 can thus be augmented nearly 200 times. Nevertheless, this technique is not more effective than those using other transfectant agents and it is difficult to implement.

Protoplast fusion

The principle consists of fusing animal cells with bacteria containing the gene to be introduced. Protoplasts are prepared beforehand from bacteria containing the recombinant plasmid. The protoplasts and animal cells are put into contact. The fusion of cell membranes, favoured by the addition of polyethylene glycol, allows the incorporation of the plasmid into the cell. This method is very rarely used.

Electroporation

Transfection by electroporation consists of a direct transfer of DNA into the cytoplasm of the host cell. Temporary pores are created in the membrane of the host cells when they are subjected to a brief electrical impulse, thus allowing the passage of the DNA. This technique offers the advantage of being usable in all types of cells. It requires special equipment, but its efficiency is clearly superior since, in some cases, nearly 90% of cells integrate the recombinant DNA in their cytoplasm. Electroporation is required for certain cell types, such as embryonic stem cells. It is also highly desirable for large fragments of DNA (e.g., BAC, YAC).

Biolistics

Miniaturization of particle gun in the form of a pistol has made it possible to transform living animal cells or tissues (see Profile 35).

Microinjection

Microinjection of genes into cell nuclei is the most effective method. It can lead to a success rate of 20% of cells in which the foreign gene is integrated. It requires special equipment and is justified only for obtaining stable clones of cells for which transfection is very difficult (e.g., some primary cells).

Transfection and apoptosis

The transfer of DNA sometimes induces apoptosis of cells, which considerably reduces the efficiency of transfection. The addition of anti-apoptosis agents can improve this situation.

Profile 37

CLONING OF ANIMALS

Cloning is by definition the reproduction of genetically identical organisms. In practice it involves a non-sexual mode of reproduction. Cloning is the natural mode of reproduction in unicellular organisms (bacteria and yeasts). In multicellular organisms, cloning involves a phenomenon of dedifferentiation. The cells of a tissue are mostly specialized in a particular function. To be specialized, they express a specific set of genes. These cells are said to be differentiated. As they become specialized, they lose their capacity to express other functions and they are frozen in their specialization. They are considered incapable of dedifferentiating. Dedifferentiation is thus the capacity of a cell belonging to a given tissue to divide and give rise to daughter cells that can produce all the cellular functions. In plants, cloning can be obtained by layering, cutting, or *in vitro* dedifferentiation of somatic cells. The cells thus dedifferentiated have the characteristics of an embryo and each can in turn result in a normal plant. No such process has ever been observed in animals.

Cloning by cleavage of an embryo

The cells of the very young embryo are considered totipotent: each cell can produce a normal embryo if it is placed in an empty pellucid zone. This operation can thus generate clones.

Mammal embryos up to the blastocyst stage can be mechanically cleaved. The two halves of the embryo can in certain conditions develop and give rise to true twins. Under natural conditions, in mammals, the birth of true twins results from an abnormal early development that produces the equivalent of an artificial cleavage of the embryo. True twins are thus clones.

Cloning by nucleus transfer

After fertilization, the nucleus of the spermatozoid, in which very few or no genes are active, decondenses to yield a pronucleus that is morphologically not much different from that of the oocyte. Some hours or days later, the genes from the spermatozoid are activated to allow the development of the embryo. The cytoplasm of the oocyte must thus have elements that allow the programming of the genome of the spermatozoid. The principle of cloning is to use this property of the cytoplasm of the oocyte to reactivate the genes of the differentiation from any type of nucleus, especially those of somatic cells rendered inactive by differentiation. The nucleus of the oocyte must be eliminated beforehand.

The mechanical transfer of the nucleus into the cytoplasm of the enucleated oocytes has thus been given rise to *Xenopus* and then sheep, cows, goats, and pigs cloned from cells that are little differentiated taken from early embryos and then totally differentiated somatic cells. The first *Xenopus* clone was obtained 40 years ago, and the first sheep 15 years ago from nuclei of non-cultured embryonic cells. The sheep, Dolly was the first animal born from a differentiated cell from an adult.

In practice, the nucleus is brought in the form of a whole cell that is placed between the pellucid zone of the enucleated oocyte and its plasma membrane. A controlled electric shock fuses the membranes and activates the new embryo thus formed by inducing calcium and other flows. In the mouse, this process has been found to be quite inefficient. The transfer of an isolated nucleus into the cytoplasm of the enucleated oocyte is necessary.

Yield of cloning

The yield of cloning has always been low. For every 100 nuclei transferred, only a few cloned animals are born. This yield is even lower when the cells from which the nuclei have been taken are more differentiated and have been kept in culture for a long time. Some cell types are easy to clone (foetal fibroblasts, cumulus cells) and others are resistant or unusable (nerve cells).

The method used to generate Dolly was based on the resting of donor cells of the nuclei by withdrawing all the growth factors from the culture medium. It was then postulated that the G_0 stage of the cell cycle (phase of cell division cycle during which the cells are at rest) was essential to obtain a clone from a differentiated cell. This hypothesis was not verified because several groups have independently obtained cloned animals with an identical yield without the donor cells having been brought first to stage G_0. The failures were caused by non-development of embryos reconstituted up to the blastocyst stage. These reimplanted blastocysts developed variably, leading often to gestations that were aborted very early. One out of two foetuses or newborns often died in the perinatal period with various but reproducible symptoms. One symptom was excess weight. The reasons for such a low yield of clones are not well known. Nevertheless, it seems to be due to epigenetic phenomena and essentially due to bad reprogramming of

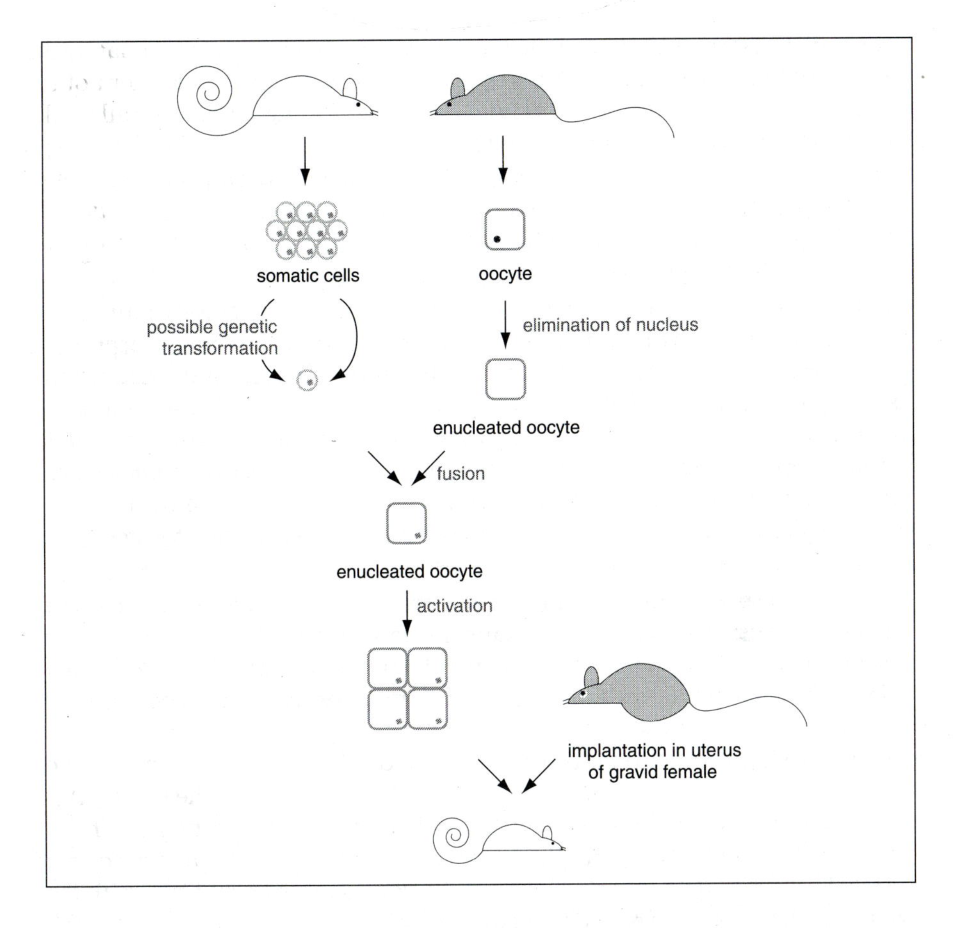

the genome. The methylation of genes, which regulates their capacity to be transcribed, is probably one of these factors.

Dolly was born with abnormally short telomeres. The telomeres of DNA are particular sequences found at the ends of chromosomes. Generally, their size is linked to the age of the cells: the older the cells, the shorter the telomeres. This preliminary observation suggests that cloning could generate prematurely aging individuals. This anomaly seemed to appear in the case of donor cells of the nucleus that themselves have short telomeres. The cells were lines obtained from an aged animal and cultured for a long period. The lambs that generated Dolly had normal telomeres. The same was true of other animals cloned subsequently. The shortening of telomeres is therefore not a feature induced by cloning.

Applications of cloning

The cloning of *Xenopus* was intended only to study phenomena of differentiation and dedifferentiation. This is still one of the major objectives of clones that are created at present. The cloning of sheep around 15 years ago was intended to accelerate genetic progress to approach what was achieved long ago with plants. The low yield of cloning presently prevents such applications on a significant scale.

Cloning, in association with techniques of genetic transformation, could simplify transgenesis in large animals. The principle is first of all to genetically transform somatic cells that donate the nuclei. Then, these transgenic nuclei are introduced in enucleated oocytes to achieve the cloning. In this way, an embryo is developed from a transgenic nucleus, and a transgenic animal is obtained. This objective has been achieved in sheep, goats, cows, pigs, and mice. Foreign genes have been added in cultured foetal cells by transfection. The cells carrying foreign genes have been used as a source of nuclei to generate transgenic cloned animals. In sheep, pigs and mice, this same approach has been used to replace genes by homologous recombination. Since the yield of cloning is found to decrease over a long period of culture needed for selection of cellular clones in which the gene has been replaced, this operation becomes difficult.

The addition and replacement of genes are presently exploited to produce recombinant proteins of pharmaceutical interest in the milk of transgenic animals, to adapt pig organs for transplants in humans, and more generally to study human diseases. Applications of cloning for livestock improvement are possible in principle. These techniques are expected to be extended to rats, chickens, rabbits, some types of fish, and other species. Reproductive cloning in humans (to obtain a cloned individual identical to the original) is presently unthinkable for strictly technical reasons, independently of ethical problems. Therapeutic cloning (applied to human cells or embryonic cells), which has been considered, could involve reconstituting an embryo from the somatic cells of a patient. Such an embryo could be the source of totipotent stem cells and then stem cells of organs obtained by differentiation of totipotent cells *in vitro*. Such an operation leads in principle to regeneration of the organ by autografting. This approach is presently theoretical and raises ethical problems that make it difficult to accept for some people, especially if the stem cells of one organ can experimentally become stem cells of another organ and thus allow the regeneration of that organ.

Profile 38

TRANSIENT EXPRESSION

In conventional techniques of genetic transformation (see Profiles 33 to 37), transformed cells are selected on a selective medium inhibiting the growth of non-transformed cells. The cells, tissues, or plants thus obtained are then analysed for the expression of the gene or genes introduced. The obtaining of transgenic organisms is undoubtedly a fundamental tool for the study of gene expression, but the procedures to obtain these organisms are often time-consuming, even difficult for some species. Transient expression can be used for rapid analysis of the expression of a gene without selection of transformed cells. Such studies can then be complemented by the analysis of these promoters in transgenic organisms (see Profile 30).

Principle

The transformed cells are cultured without selection pressure, harvested 24 to 72 hours after transfection (period during which the genes carried by the plasmid are transcribed and translated before their integration into chromosomes), then ground in order to study the expression of the gene introduced. In general, genes known as reporters, the expression of which can easily be detected, are used in these experiments.

Definition of a reporter gene and examples

The product of a reporter gene must be absent from cells subjected to the transformation. It is for this reason that, with respect to eukaryotic cells, the reporter genes are mostly of bacterial origin. The protein coded by the reporter gene must be easily detectable; it is often an enzyme the expression of which can be visualized by simple staining or measured by an enzymatic activity test. The reporter genes most commonly used are the following:

- The gene for chloramphenicol acetyl transferase (CAT). The activity of CAT, catalysing the transfer of an acetyl group on chloramphenicol, is easy to measure in the presence of chloramphenicol and acetyl-coenzyme A labelled with ^{14}C.

- The gene *lacZ* of β-galactosidase[1] and the gene *gus* of β-glucuronidase. The expression of β-galactosidase or β-glucuronidase is detectable by blue staining of cells in the presence respectively of X-Gal or X-Gluc.
- The gene for luciferase. In the presence of luciferin and ATP, luciferase expressed in cells is capable of inducing light emission that is easily located and measured.
- The gene for green fluorescent protein (GFP). The GFP expressed emits a green light (at 509 nm) that can be seen by fluorescence under UV light. The fluorescence is intrinsic to the protein and its detection requires no infiltration of tissues by a substrate because there is no enzyme. This reporter gene can thus be used for *in vivo* experiments.

The choice of reporter gene depends on the type of tissue studied and the intended use. For example, the *gus* and *gfp* genes are appropriate for *in situ* location of activities of promoters studied. The *cat* gene is more suitable for quantitative measures of the expression from raw cell extracts.

[1]β-galactosidase breaks the glycoside bond of its substrate, X-Gal. The chromophoric group X is thus liberated, which induces a blue coloration. The same is true of β-glucuronidase, which catalyses the breaking of the glucuronide bond of its substrate, X-Gluc.

VII

Analysis of Gene Function

Profile 39

RECOMBINANT PROTEINS

The study of proteins involves their large-scale production and purification. The strategy developed here is to use bacteria (generally *E. coli*[1]) to mass produce the target protein in the form of a fusion protein: the fusion protein is made up of a fragment of bacterial protein with which the desired eukaryote protein (protein X) is associated by the intermediary of a short sequence. The fusion protein, often more stable than protein X alone, can be produced in large quantity.

Principle

The gene coding for the target protein X is cloned in a vector of expression. This vector is a plasmid with a bacterial promoter (most often inducible) followed by part of the sequence coding for a bacterial protein, of a sequence corresponding to a cleavage site recognized by a protease and a polycloning site at the level of which the target gene will be introduced, respecting the reading frame for translation into amino acids. After transformation, the bacteria that have integrated a recombinant vector are selected on a selection medium, multiplied, then harvested and lysed so as to obtain a raw cellular extract containing all the proteins synthesized, including the fusion protein. From this raw extract, the fusion protein is purified by affinity chromatography on column: groups, called ligands, are fixed on the column and interact specifically with the fusion protein, retaining the latter, while the other proteins are eliminated. The fusion protein is recovered after specific elution and a cleavage is induced between the bacterial protein and protein X by incubation in the presence of a suitable protease. A second affinity chromatography can be carried out to eliminate the bacterial protein: only the latter is retained on the column, while protein X is eluted.

Many systems are now available. The choice of one over the other must take into account various parameters such as the stability of the fusion protein synthesized, its solubility, or the conditions of elution. Some examples (not an exhaustive list) are given in the table that follows. New ligands are regularly proposed. For example, one type of fusion that is increasingly commonly used consists in grafting the cDNA coding for a heat-stable protein (thioredoxin) to the DNA sequence that is to be translated into amino acids: the addition of the thioredoxin can in most cases increase the solubility of the protein desired and/or its thermal stability.

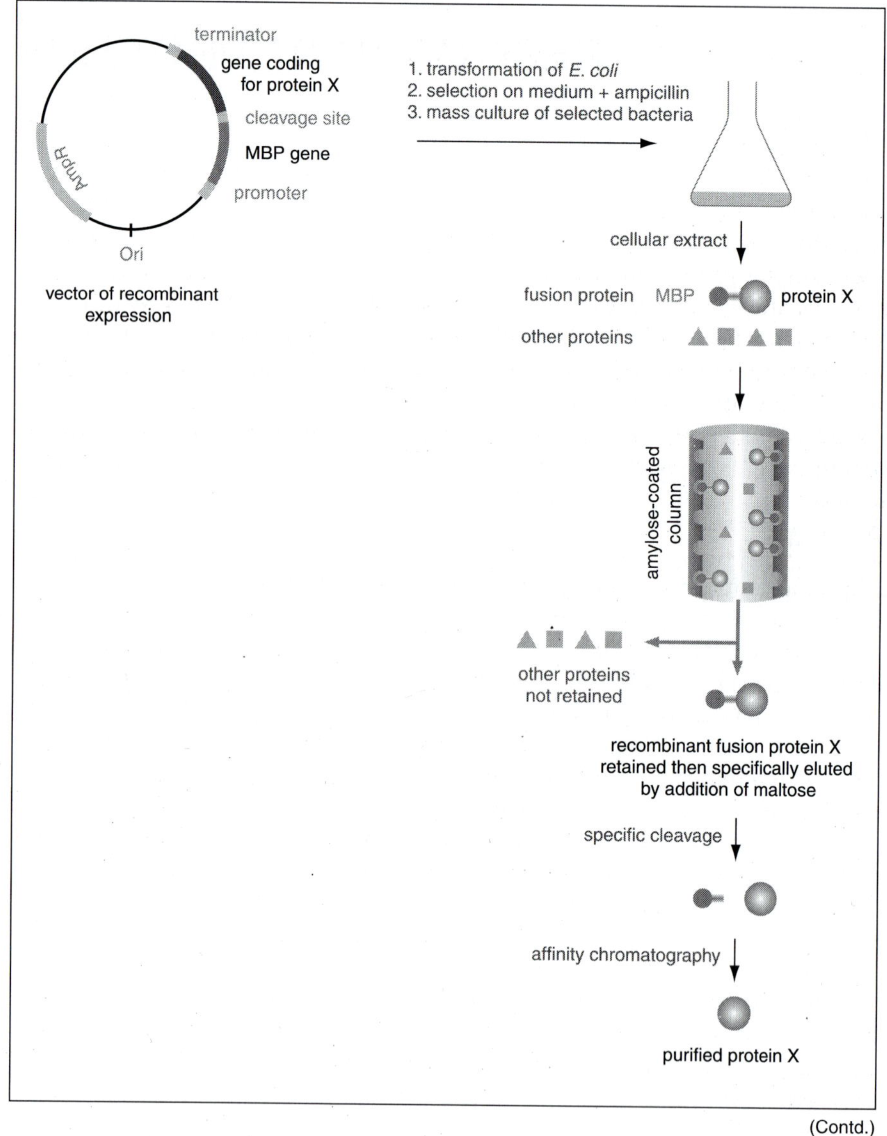

(Contd.)

Bacterial peptide or fusion protein	*Ligand*	*Agent permitting specific elution*
maltose binding protein (MBP)	amylose	maltose
glutathione S-transferase	glutathione	reducing agent
6 histidines (His-Tag)	Ni^{++} or Zn^{++}	imidazole, pH gradient
calmoduline binding protein (CBP)	calmoduline	EGTA
ZZ domain[2]	IgG	low pH
intein	chitin	reducing agent

Recombinant non-fusion proteins

Even though the technology of producing recombinant protein in the form of fusion proteins considerably simplifies the task of purification, it may sometimes (for crystallography and RMN studies) be desirable to obtain the recombinant product in the non-fusion form and in large quantity. An alternative to production of fusion proteins is the use of plasmids of the pET series, where the promoter is that of T7-RNA polymerase from a virus that confers a speed of transcription much greater than that of promoters of *E. coli*. In consequence, the cDNAs placed under the control of this type of promoter are transcribed at very high rate and the corresponding protein is produced in very large quantity (up to 50% of the total proteins of *E. coli*). This generally allows easy purification of the protein, which is very slightly contaminated by possible proteins of homologous function present in *E. coli*. However, large-scale production of a foreign protein may be harmful to the bacterium (which will not grow). Moreover, in the absence of the fusion protein, purification of the target protein remains more delicate and requires traditional techniques of protein biochemistry.

[1]Recombinant proteins can be produced in eukaryotic cells such as yeast or insect cells (see Profile 40).

[2]The ZZ domain is the domain of fixation of IgG of protein A.

Profile **40**

BACULOVIRUSES OF INSECTS, VECTORS OF EXPRESSION OF FOREIGN GENES

The baculoviruses (from the Latin *baculum*, rod) constitute a vast group of viruses. They are present only in the Arthropods and have been described in more than 600 insect species as well as in some crustaceans. These are viruses packaged in the form of rods, about 350 nm long with an average diameter of 50 nm. They contain a genome made up of a double-stranded, circular, tightly rolled DNA molecule varying in size from 88 to 153 kbp.

One of the most thoroughly studied baculoviruses at present is the virus that was first isolated from larvae of the butterfly *Autographa californica*. This virus, known as AcMNPV (*Autographa californica* multiple nuclear polyhedrosis virus), multiplies in the cell nuclei of more than 30 species of lepidopterans. In this case, several viral particles are contained in a single package (in other cases, there is only one particle per package).

In nature, viruses are present inside polyhedrons, crystalline structures easily observed under light microscope that form virtual "inclusion bodies". The protein matrix of these inclusion bodies is made up of polyhedrin, a protein coded by the viral genome. The virus, protected from environmental factors by this body, is disseminated when the insect dies. After the polyhedrons are ingested by another insect, they are dissolved in the intestinal juices, releasing the viral particles, which can penetrate the cells of the intestinal epithelium by membrane fusion or by endocytosis. After the viruses go into the intestinal epithelium, they infect the host tissues. In the cells, the nucleocapsid is brought to the nucleus, where it releases its DNA. The transcription of early viral genes then begins. The replication of the DNA begins around 7 hours after the infection and the expression of early genes is repressed in favour of that of the later genes. After 10 hours, viral

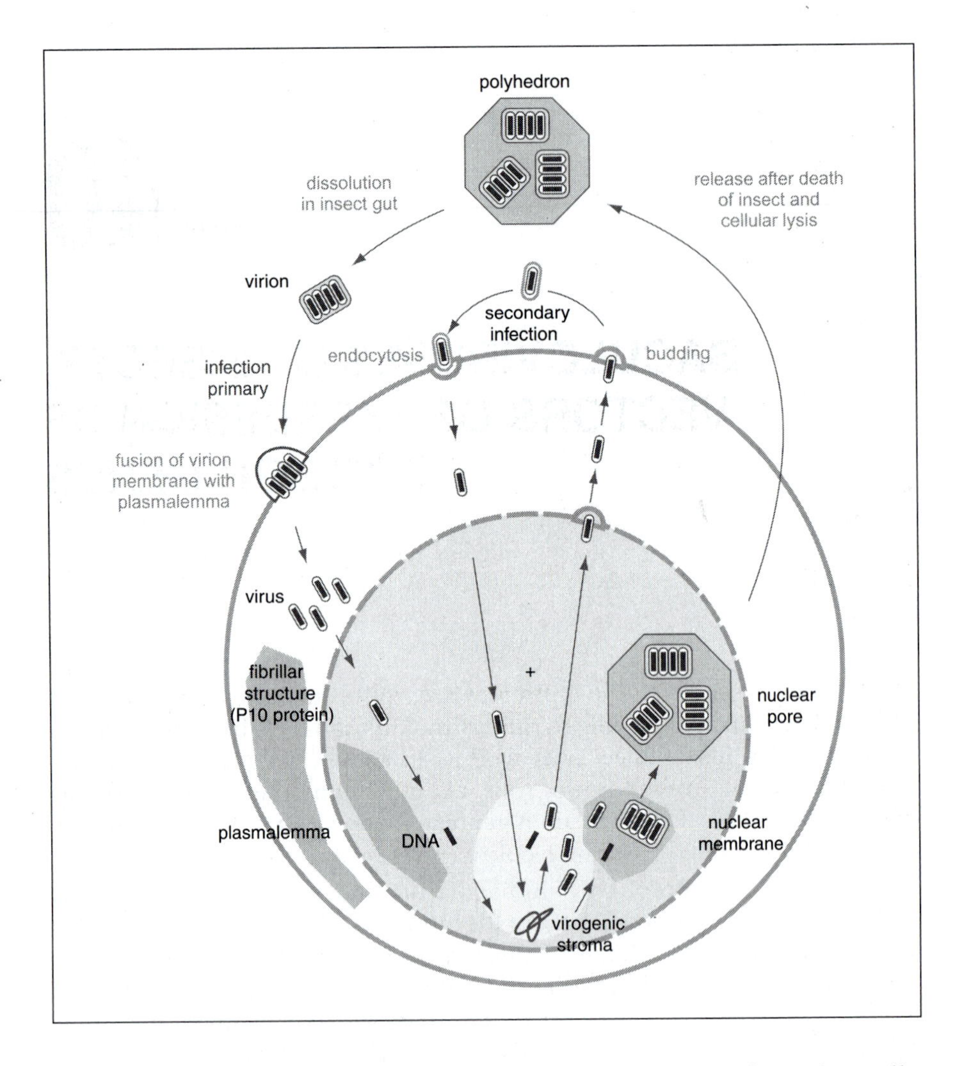

particles bud at the cell surface and will be released to infect other cells. Later, in the last stage end of the infection, at least two genes (polyhedrin and P10 protein, the latter forming fibrillary structures in the nucleus and cytoplasm) are expressed in a very large quantity. The role of P10 is still not clear; it is involved in cellular lysis. On the other hand, polyhedrin is essential to the formation of inclusion bodies, which make it possible for the virus to spread in the wild. However, in *in vitro* cell culture, the polyhedrons are not infectious and, under these conditions, the expression of the polyhedrin gene, like that of the P10 gene, is not indispensable to obtaining a cycle of multiplication.

Principle

The principle of the use of baculoviruses as vectors of foreign genes is simple: the region of the genome that codes for polyhedrin and/or P10 protein is replaced by a DNA fragment that codes for a target foreign protein. To date, more than 3000 genes of varied origin have been expressed by this means. One advantage of using baculoviruses is that some post-translation modifications of foreign eukaryotic proteins can be carried out (see later). Recombinant baculoviruses expressing a foreign gene are generally constructed in two steps. The foreign gene is inserted in a bacterial plasmid that contains a single cloning site downstream of the promoters of the polyhedrin (*pol*) or P10 gene surrounded by flanking sequences of DNA of the baculovirus: this is the transfer vector. Then cultures of insect cells are co-transfected with the transfer vector "loaded" with the heterologous gene and the DNA of the wild virus that is infectious. By this means, a recombination can be produced between the homologous regions of the plasmid and of the viral genome, which results in the obtaining of a recombinant virus that has integrated the foreign gene.

Recombinant baculoviruses are often identified by the appearance of layers of infected cells that form lysis plates, visible under the microscope with slight enlargement. In such a cell layer, one infected cell stops dividing and rapidly produces budding viral particles. These will infect the neighbouring cells. To prevent the dispersal of these virions, a layer of agarose is placed on the cells. After 4 to 5 days, many cells that surround the initial cell are also infected and die. Thus, by using a bright colour such as neutral red, it is very easy to distinguish these lysis plates, which form colourless "holes" in the red layer of living cells. All the plates examined under microscope will present either a "polyhedron presence" phenotype corresponding to infection by a wild virus that has not been recombined or a "polyhedron absence" phenotype corresponding to a recombinant virus that has the foreign gene in place of the polyhedrin gene. These plates can thus easily be recovered and subcultured once or twice to ensure the purity of the recombinant virus. Several independent recombinants are recovered, purified, and tested for the expression of the recombinant protein. In this way, the foreign protein can be produced and studied. It is obvious that either the polyhedrin promoter or the P10 promoter can be used, or both, which would then allow the expression of proteins made up of several subunits such as immunoglobulins.

Post-translation modifications of proteins produced in insect cells

The proteins produced in insect cells, in the large majority of cases, present biological activities similar to those of native proteins: e.g., enzymatic, antigenicity, immunogenicity, and other activity.

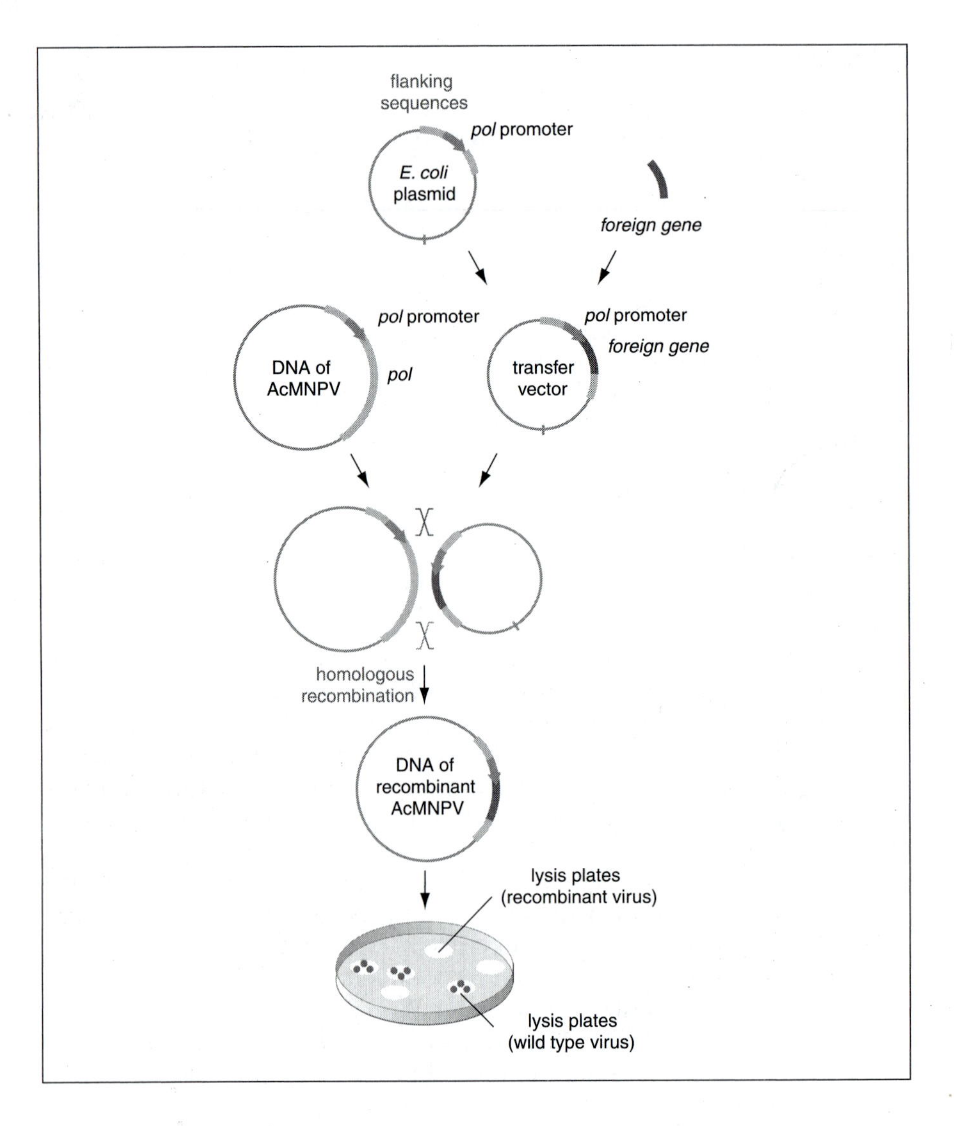

Insect cells have enzymatic equipment that allows them to carry out post-translation modifications similar to those of vertebrates. They recognize homologous and heterologous signal peptides, and mature proteins are excreted into the culture medium. In the same way, the addressing signals to different cell compartments are recognized: nuclear, cytoplasmic, or membrane targeting. Polyproteins are most often cleaved in the same way as the native protein. The proteins produced are correctly phosphorylated, myristylated, palmistylated, farnesylated, carboxymethylated, amidated, or acetylated. The assembly into dimers, in an oligomeric complex, and the formation of disulphide bonds occur in a correct manner.

One important point concerns the glycosylation of proteins. A detailed analysis of glycoproteins expressed in insect cells has shown differences in the size and structure of oligosaccharides in relation to those of glycoproteins of mammals or plants. The glycoproteins of mammals expressed in the cells of *Spodoptera frugiperda* (Sf9) are often smaller and the oligosaccharide chains are shorter. In a large number of cases, this difference does not affect, or only slightly affects, the biological activity of proteins *in vitro*. This problem of glycosylation seems even now one of the only post-translation modifications that cells of *S. frugiperda* do not carry out in a manner identical to that of mammal cells.

Culture of insect cells

A large number of insect cell lines are available nowadays to study the multiplication of the baculovirus of *A. californica*: the most commonly used are Sf21 cells and the clone Sf9, which come from the ovarian tissue of the butterfly nymph. The TN-368 line, derived from *Trichoplusia ni*, as well as the Mb line, derived from *Mamestra brassicae*, also result in a very high multiplication of the baculovirus.

Most lepidopteran cells have an optimal temperature for their growth between 5 and 30°C. Unlike mammal cells, they do not require the use of an enriched-CO_2 atmosphere and the addition of phenol red. A buffer effective at pH 6.2 is sufficient. The time required for the cell population to double is 16 to 24 hours and under normal conditions the cell density varies from 2×10^6 to 5×10^6 cell/ml. This density can be increased in a bioreactor. Finally, most insect cell lines can grow in a cell layer or in suspension. The culture of insect cells in large volume bioreactors is now increasingly better controlled and bioreactors of 5 to 100 l are frequently found in pilot installations.

Profile **41**

YEAST TWO HYBRID SYSTEM

The two hybrid system was developed to reveal protein-protein interactions. This test is based on the fact that many proteins, including transcription factors, are composed of several independent functional domains. When these domains are physically distant, they have no biological activity. But if they are brought close to one another by non-covalent interactions, a functional protein can be reconstructed. The Gal4 protein activator of the transcription of genes for galactose metabolism in the yeast *Saccharomyces cerevisiae* is composed of two distinct and separable functional domains: a DNA binding domain (DBD) and a transcription activation domain (AD).

Principle

Suppose that there are two proteins X and Y for which a possible interaction is to be tested. Each of the two domains of the Gal4 protein can be fused separately with each of the target proteins. Thus, protein X will be linked to the DBD and protein Y to the AD. If proteins X and Y interact, the DBD and AD of the Gal4 protein will come together and an active Gal4 protein will thus be reconstituted. Since the biological activity of the Gal4 protein is the transcriptional activation of genes, the reconstructed Gal4 protein could activate the transcription of a reporter gene (see Profile 38).

In order to carry out the test, the genes X and Y are cloned downstream of sequences coding for the DBD or AD of the Gal4 protein. These two plasmids are used to transform a yeast strain that contains the reporter gene *LacZ* coding for β-galactosidase under the control of a promoter that can be activated by the Gal4 protein (Gal1 promoter). The reporter gene will be transcribed only if the two proteins X and Y, interacting together, reconstruct a functional Gal4 protein. The product of the reporter gene *LacZ* can thus be detected by measurement of its enzymatic activity or by blue staining of yeast colonies cultivated on agarose medium in the presence of a substrate that will give a coloured reaction (X-Gal, see Profile 38).

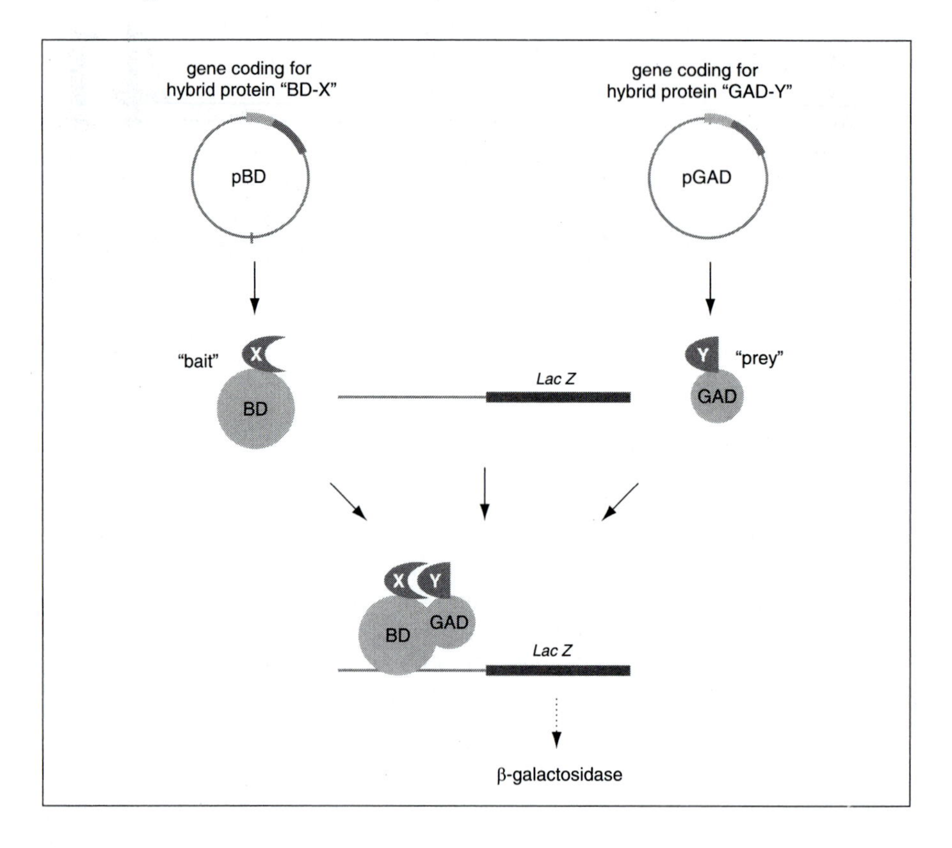

Applications

This system makes it possible to study simply and economically the interactions between proteins from any eukaryotic organism. However, the Gal4 protein is active in the nucleus, since it fixes itself on the DNA. Thus, the double hybrid test is useful only if the interaction between the two proteins X and Y can take place in the nucleus of the cell. Various strategies have thus been developed to study interactions between two membrane proteins or two cytosolic proteins.

Membrane proteins

In the example presented here, an attempt was made to find out whether protein X (the bait) interacts with protein Y (the prey) at the membrane level. To carry out this assay, it is necessary to work with a mutant yeast strain that does not express the GEF (guanine nucleotide exchange factor) protein. In the absence of GEF, the Ras protein is not activated and the yeast cannot grow.[1] The gene for protein X is fused with a DNA segment coding for a myristylation signal (Myr) that makes it possible to locate it on the membrane. The yeast expressing this gene will thus have the protein X anchored in the plasmalemma near the cytosol. The gene for protein Y is fused with that of a protein called SOS (the equivalent of GEF in mammals). The interaction of X and Y allows the location of the SOS protein on the membrane. SOS thus activates the Ras protein (by exchanging its GDP for a GTP) and the yeast can grow at a restrictive temperature, while in the absence of the interaction between X and Y, no growth is possible, and thus no colony is visible.

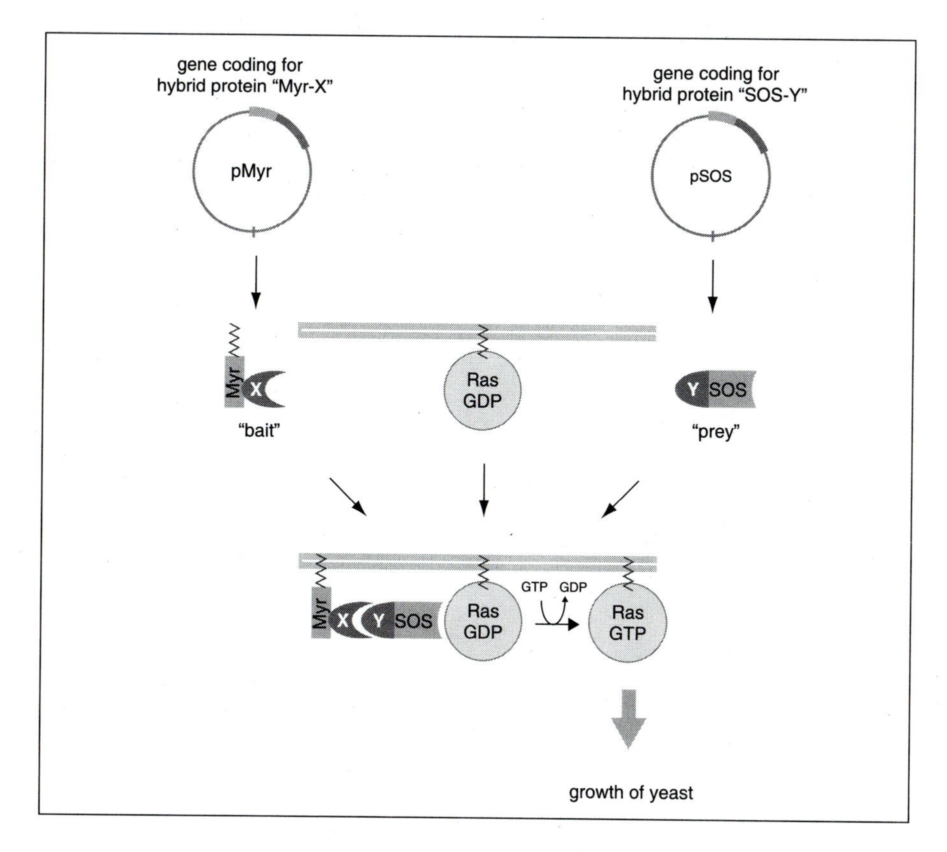

Cytosolic proteins

In eukaryotic cells, the proteins fixed at the ubiquitin peptide are recognized and degraded in the cytosol by the protein complex of proteasome. To find out whether proteins X and Y interact in cytosol, the gene for protein X is fused with a fragment of a gene for ubiquitin coding for the N-terminal part of this peptide (labelled N in the figure), which is itself fused with a gene coding for a transcription factor (labelled S) that controls a reporter gene. In parallel, the gene for protein Y is fused with a fragment of the gene for ubiquitin coding for the C-terminal part (labelled C). If X and Y interact, the ubiquitin is reconstructed and the system is recognized by the proteasome. The cleavage obtained following the interaction between X and Y allows the release of the transcription factor, which in turn activates a reporter gene in the nucleus, which generates a signal. Consequently, the interaction between X and Y could be detected by selection of colonies by a colorimetric test.

The two hybrid system is also a highly effective technique to find not only a partner, but all the partners that could interact with the protein of interest X. In this case, a cDNA expression library is constructed (see Profile 15) in a vector that allows the fusion of the AD at each cDNA of the library. The known protein X is fused to the DBD. The cDNA library is screened by incubation of bacteria expressing the protein Y fused with AD with the protein X fused with DBD. However, this technique still generates a significant number of "false positives".

Interactome

The sequencing of complete genomes has radically changed the search for cellular partners by the two hybrid system. In this case, a catalogue of all the

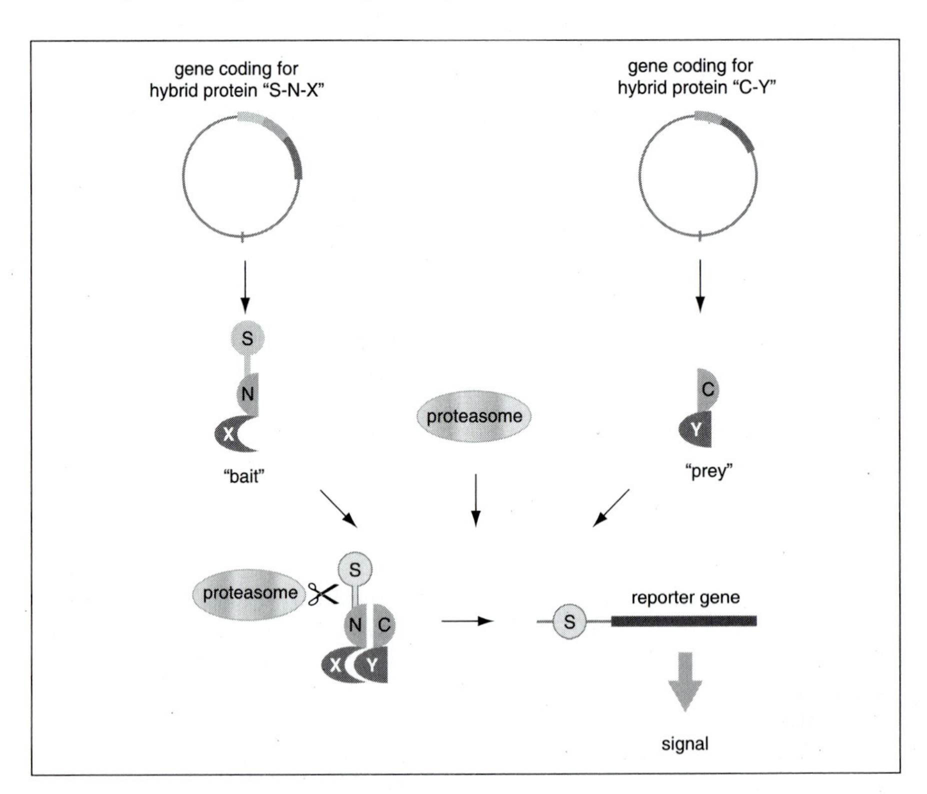

possible protein-protein interactions can be compiled. Yeast contains around 6000 genes. In order to map the interactome of yeast, each of these 6000 genes must be fused with the AD and a collection of 6000 yeasts, each containing a different construction, will be obtained. Meanwhile, an identical collection is constructed by fusing each of the 6000 genes with DBD. The two collections must be constructed on haploid and sexually compatible yeast strains. All the 2×2 crosses are performed and the results analysed. This approach requires large-scale automation and statistical analysis of the matrixes obtained. A great deal of work remains to be done to achieve an interactome of the human genome (which has more than 30,000 genes).

---

[1]The mutation is heat-stable, which allows the mutant strains to be conserved at a permissive temperature for which the mutation is silent.

Profile 42

SITE-DIRECTED MUTAGENESIS

Site-directed mutagenesis is a technique that allows the *in vitro* modification of the sequence of amino acids of a protein in order to evaluate the importance of the modified amino acids in the functioning of the protein. The principle is to mutate the gene *in vitro* and to produce the protein *in vitro*. In order to do this, it is necessary to have the cDNA or the coding sequence of this protein. The nucleotide sequence, modified *in vitro*, is reintroduced in a microorganism (generally a bacterium, see Profiles 10 and 39). The transformed microorganism synthesizes the protein from a mutant gene to produce a polypeptide harbouring the mutation. The biochemical characteristics of this mutated protein are then analysed in order to study the importance of mutated amino acids in the functioning of the protein. Three types of mutations can occur: deletion, insertion, or substitution of bases.

Principle

Mutations by deletion or insertion can be obtained easily if the gene sequence contains appropriate cleavage sites. By digestion using adequate restriction enzymes, it is thus possible to eliminate part of the sequence or to introduce a fragment of foreign DNA at a definite position. Nevertheless, this approach is limited because it does not allow mutation targeted at a small number of amino acids. To overcome these disadvantages, a technique of site-directed mutagenesis *in vitro* in the presence of mutagenic oligonucleotides of known sequence has been developed. This method allows modifications to be made at a selected position in a DNA fragment of known sequence. For example, it is possible to change a single amino acid of a protein for which the corresponding gene sequence is known.

The basic principle lies in the synthesis of an oligonucleotide sequence complementary to that of the region of the gene to be mutated, but possessing the desired mutation. The gene to be mutated is introduced in a phagec vector of M13 type and the DNA is purified in the single-stranded form. The mutagenic oligonucleotide is then hybridized with the target single-stranded DNA. The double strand thus formed is used as a primer by polymerase DNA, which synthesizes the complementary strand. The molecules obtained correspond to a double-stranded plasmid, one of which possesses the mutation and the other does not. After transformation of bacteria, some will process the mutant strand, others the wild type. Sequencing of a small number of clones is thus necessary to identify the phages possessing the mutant sequence. The mutant proteins are then produced in *E. coli* (see Profile 39) in order to analyse the effect of the modification of the modified sequence.

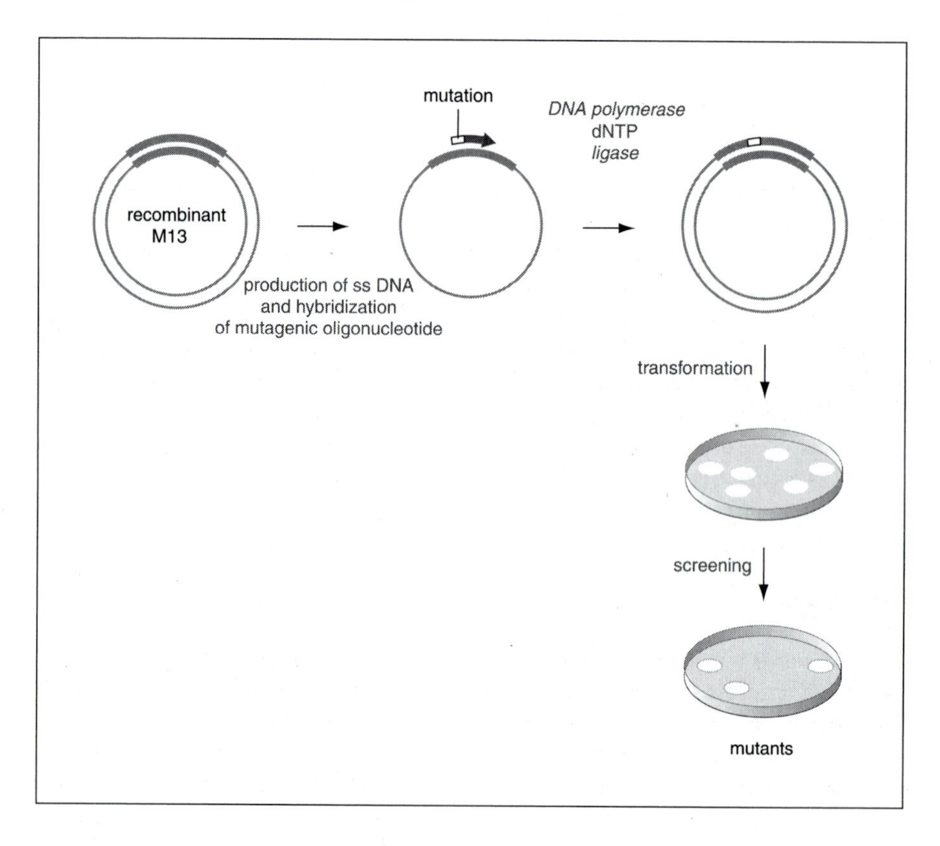

PCR-directed mutagenesis

Since the yield of mutagenesis described in the preceding paragraph is relatively low, some modifications have been made to this basic principle. The introduction of a mutation by PCR makes it possible to create 100% mutants, whereas the preceding technique (called the Kunkel technique) results in only 80% mutants. The principle of a mutation introduced by PCR is based on the fact that strict complementarity of the nucleotide primer with the target DNA sequence is not absolutely necessary throughout the length under consideration: it is essential on the 3'-OH end for the priming of the reaction but not on the 5'-P end.

The first method of PCR-directed mutagenesis requires the use of four different oligonucleotide primers. Two oligonucleotide primers (denoted 1 and 2) are complementary to each other at the level of the mutant 5'-P region. The other two (denoted 3 and 4) allow the amplification of all the cDNAs that needs to be mutated. They are thus complementary to the 3'-OH and 5'-P ends of the cDNA present in the plasmid.

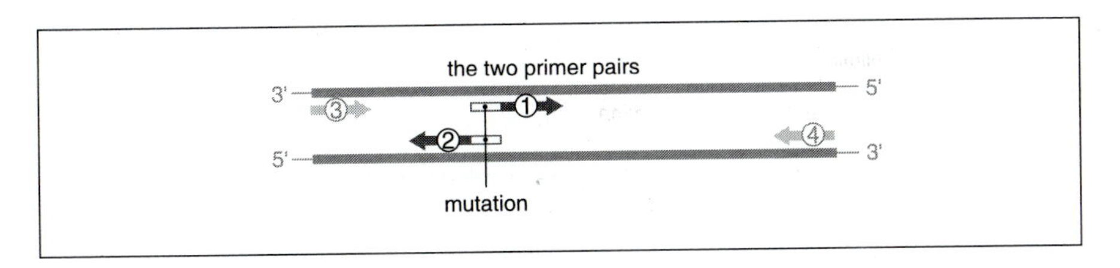

Using these two pairs of primers, three successive PCR reactions are carried out. During the first PCR, the use of primers 1 and 4 makes it possible to obtain a partial cDNA fragment mutated on the desired region. In parallel, the second PCR is carried out using primers 2 and 3 to obtain a partial and mutated cDNA fragment in the same way. Combining these two amplifications and denaturing the DNA leads to a renaturation between the fragments at the level of modified central primers. This hybridization causes 3'-OH ends to appear, from which the fragment can be elongated up to the ends of the target sequence. The continuation of cycles in the presence of primers 3 and 4 makes it possible to obtain a complete fragment carrying the mutation on the two strands. This fragment can then be sub-cloned in a vector for the production of the protein.

The second method of PCR-directed mutagenesis is called the "megaprimer" method. It requires the use of three oligonucleotide primers. The first carries the mutation that is to be introduced (denoted 1) and is complementary to one of the strands of the target DNA. The two others (denoted 3 and 4), as before, correspond to the ends of the cDNA.

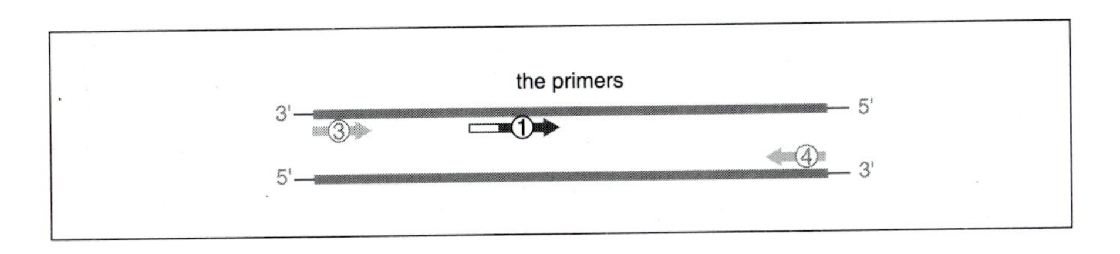

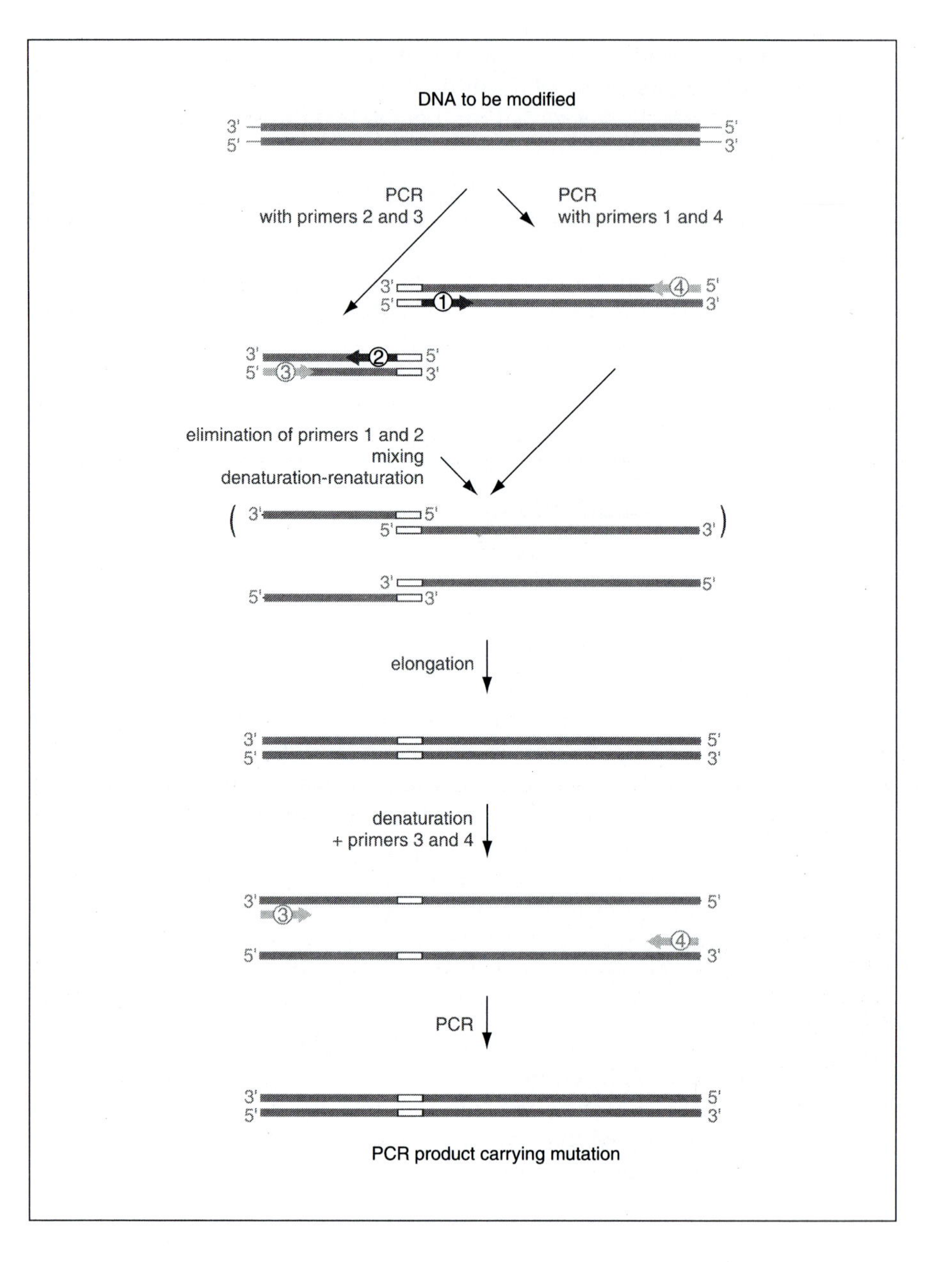

In the first PCR reaction, primers 1 and 4 can be used to obtain a PCR fragment carrying the mutation and corresponding to a truncated sequence of the matrix cDNA. This fragment is purified and will serve as a primer (megaprimer) in a second PCR reaction with oligonucleotide primer 3. In this fashion, a complete and mutated fragment is generated.

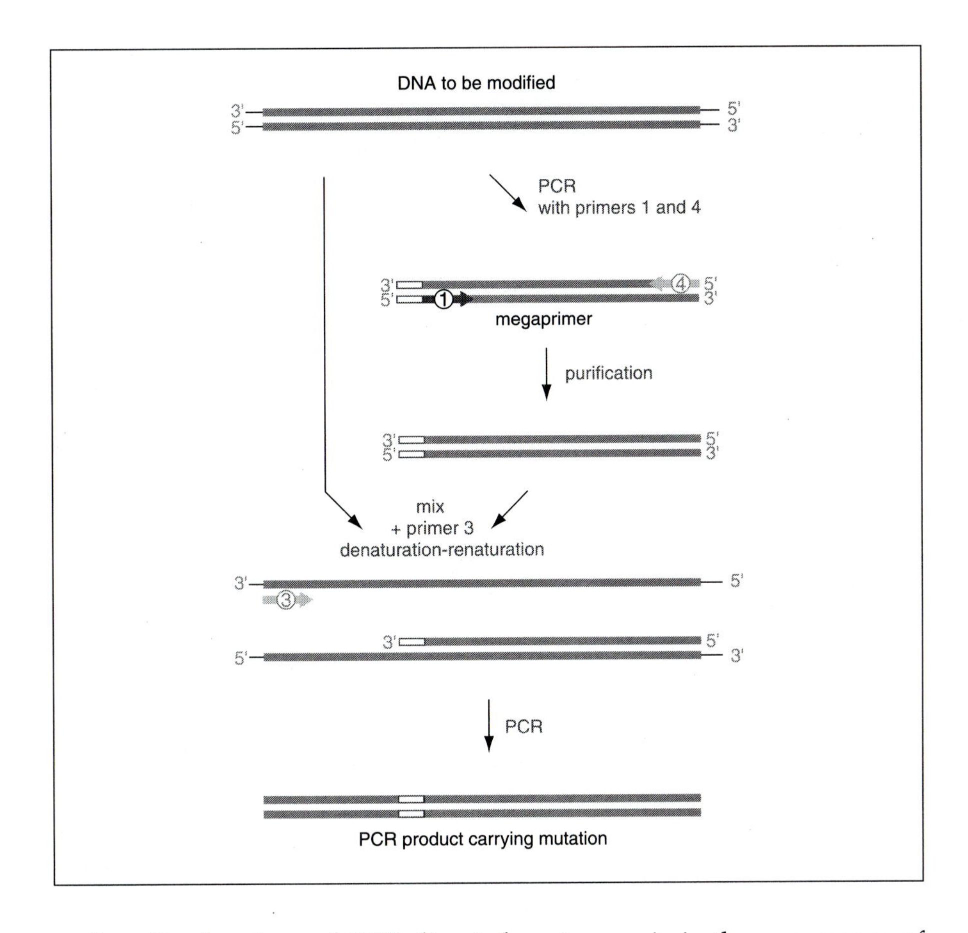

One disadvantage of PCR-directed mutagenesis is the occurrence of errors in reading by *Taq* polymerase. This problem is easily overcome since a reduced number of cycles is usually used (<30) and because new generations of thermo resistant DNA polymerases reputed to be error-proof are now commercially available.

Profile **43**

MUTATION COMPLEMENTATION IN YEAST

Many genome sequencing projects result in the characterization of genes of unknown function or genes identified by simple homology of sequence. One way to demonstrate the biological function of these genes is to complement a mutation of the function studied in yeast. Genetic transformation can be used to integrate the target gene in the mutant yeast that has lost a known function, then the phenotype of the transformed yeast is determined in order to discover whether the introduced gene is capable of recovering the lost function or, in other words, complementing the mutation.

Many yeast proteins have a homologue in mammals or other eukaryotes. It is thus possible to study the function of a gene of another eukaryote in yeast. This technique is ideal for the characterization of genes that code for homologous proteins of *S. cerevisiae* involved in the transport of metabolites or ions, in the secretion of proteins, or in the control of the cell cycle. For this purpose, a complete collection of deleted yeast strains is now available (URL: http://www.uni-frankfurt.de/fb15/mikro/euroscarf). These mutants were obtained by insertion of a deletion cassette (containing a kanamycin resistance gene) at the chromosome level. In the case of genes belonging to multigenic families (e.g., hexose transporters), it may be necessary to delete several genes before obtaining a mutant phenotype. These mutants carrying multiple deletions can then be used for complementation by means of heterologous genes belonging to multigenic families. Some of these mutations confer a specific phenotype on the mutant strain, which differentiates it from the parent strain (e.g., auxotrophy with respect to an amino acid when an enzyme of its biosynthesis pathway is deleted, absence of growth in a low concentration of nutrients when the transporters involved are deleted). Other mutants are called conditional, that is, the mutant gene remains functional under certain conditions and not under others (e.g., heat-sensitive mutants that can grow at a permissible temperature but not at a higher temperature).

The principle consists in isolating a yeast mutant carrying a useful deficiency. The mutant is then transformed using a target sequence. This sequence must be placed under the transcriptional control of a promoter and a terminator that come from a yeast gene. Moreover, the plasmid used containing the transgene must carry a selection gene for the yeast, for example the gene *His3*.[1] After transformation, the cells are spread out in the petri dish under conditions that allow only the development of yeast transformed by the plasmid. Subcultures are then carried out on nylon membrane and used to select the strains containing the wild type gene that allows the deficient activity to be restored. The plasmid contained in such a strain is isolated and analysed.

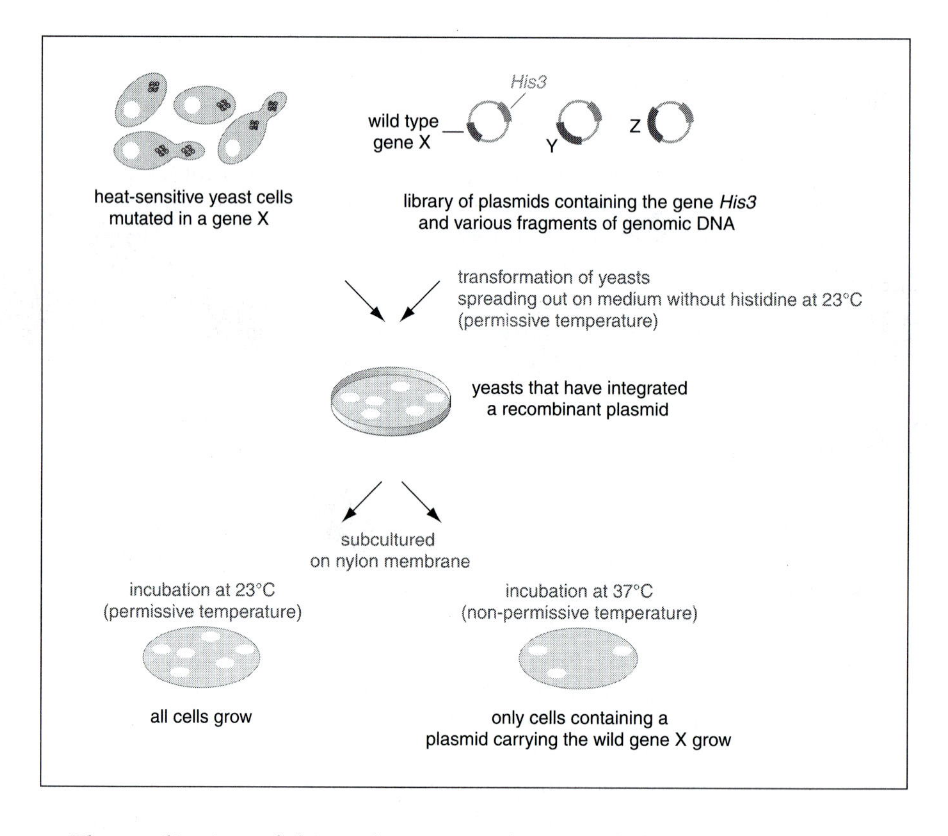

The application of this technique can be extended to a search for a gene that has a particular function within a given genome, if there is a yeast mutant for that function. In that case, a cDNA or genome library is constructed in a vector allowing the expression of these sequences in yeast (same type of plasmid as that used earlier). The target mutant is then transformed by means of the entire DNA library and the complemented cells are selected. Once the transformations with the appropriate phenotype are obtained, it is essential to confirm that the complementation is linked to the plasmid. In fact, the complementation of a deficiency can sometimes be due to a suppressor, the overexpression of which can counteract the effect of the mutation. It is necessary to test for loss of the plasmid, by growing the strain on a rich medium, for example. The yeasts that have lost the plasmid must be incapable of developing on the selection medium. After this test, the plasmid carrying the target gene can be isolated from the yeast strain, multiplied in *E. coli*, and reintroduced in the deficient yeast in order to confirm the complementation of the deficiency.

[1]The gene *His3* codes for imidazole glycerol phosphate dehydrogenase. It allows the selection of yeasts expressing this gene by growth on a medium without histidine.

Profile **44**

KNOCK-OUT OF GENES IN YEAST

The characterization of a gene by cloning and sequencing leads to comparison of its sequence with those deposited in international data banks. In some cases, the gene identified presents similarities with a gene of another organism coding for a protein of known function. In other cases, the sequence studied has no similarity with any known sequence and the function of the protein coded by that gene remains unknown. In these two types of situations, it remains necessary either to demonstrate the biological function identified by similarity experimentally or to create an individual deficient in the function of the unidentified gene. In other words, the role of the gene must be confirmed or determined. The gene knock-out technique makes it possible to obtain a transgenic organism that no longer expresses the target gene. The phenotype of this mutant, which lacks a function, will thus indicate the role of the protein coded by the gene that is being studied.

This technique involves genetic transformation of an organism (see Profiles 33 to 37) in the course of which a transgene is integrated in the sequence of the target gene and not randomly in the genome. To do this, restriction enzymes (see Profile 4) are used to modify a target gene *in vitro* by placing the sequence coding for a selection gene in it. In this way, the translation of a target gene will give an incomplete and inactive protein. This construction is thus integrated in the organism by transformation. The transformed cells are selected on the medium in the presence of selection agent. During the transformation, the transgene may either integrate itself randomly in the genome or integrate itself by homologous recombination at the expected locus. Therefore, cells that have undergone the homologous recombination must be tested. In yeast and in most fungi, selection is done after transformation, by analysing the site of integration of the transgene in the genome of cells resistant to the selection agent. It is sufficient for the genomic DNA to be cleaved by restriction enzymes, which will make it possible to discriminate between the endogenous gene that is not affected by the insertion and the endogenous gene modified by the transgene. The organisms thus obtained become incapable of synthesizing the target protein and their phenotype can be studied. The homologous recombination, and thus the gene knock-out technique, are not applicable to all eukaryotic organisms. The technique remains to be proven, especially in plants.

During a homologous recombination, two recombination events, one at each end of the gene, are needed to replace the chromosomal gene precisely with the mutant gene. It is a rare event and the frequency of homologous recombination is low. Moreover, the inactivation of a gene ends in a recessive mutation (loss of function) that is then silent in the diploid individuals: the event of homologous recombination affects only one of the two alleles on the locus involved, the other being able to express and yield an active wild type protein. It is thus necessary to isolate haploid individuals carrying the mutation. In yeast, for example, sporulation is induced that produces by meiosis four haploid spores (equivalent to gametes): two of them carry the inactive mutant gene as well as the marker gene, and two of them have the active wild type gene and lack the marker gene. These spores are allowed to germinate on a selection medium in order to select strains descended from transformed spores.

This technique can also be applied to the animal model using embryo cells of mammals and a modified gene carrying an adequate selection marker. In this case, some heterozygous transgenic animals are obtained in the first set of descendants. They must be crossed among themselves to obtain homozygous individuals the phenotype of which will be studied. However, the frequency of the homologous recombination remains very low: in mouse, for example, the frequence is one-hundredth the frequency in yeast.

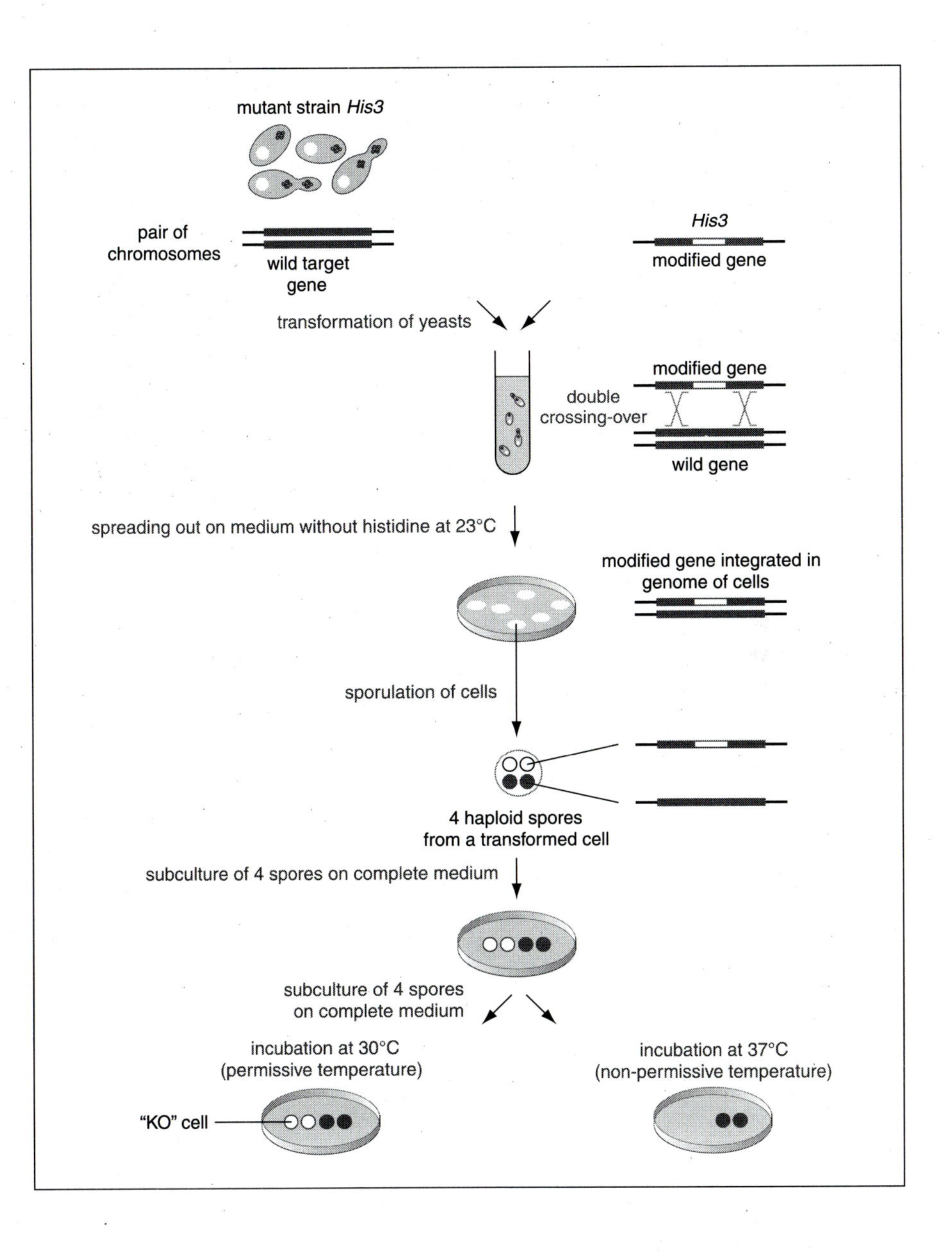
mutant strain His3
pair of chromosomes
wild target gene
His3
modified gene
transformation of yeasts
modified gene
double crossing-over
wild gene
spreading out on medium without histidine at 23°C
modified gene integrated in genome of cells
sporulation of cells
4 haploid spores from a transformed cell
subculture of 4 spores on complete medium
subculture of 4 spores on complete medium
incubation at 30°C (permissive temperature)
incubation at 37°C (non-permissive temperature)
"KO" cell

Profile **45**

GENE TAGGING

The general term "gene tagging" encompasses a set of strategies that are used to establish a link between the sequence of a gene and its function, upstream of conventional studies of expression and protein activity.

The basic principle of gene tagging is the use of a foreign DNA sequence capable of inserting itself into different places in the genome. By inserting itself in a given gene or close to it, the foreign sequence will (1) affect the function of this gene and create a mutation in it; (2) serve as tag to allow the gene in question to be materially located and to establish the sequence-function link in short. The integration of the foreign DNA sequence requires a transgenesis step and the establishment of a collection of transformed or mutant individuals.

Two approaches are possible: (1) conventional tagging or forward genetics, which allows the cloning of a gene from a mutant phenotype, or (2) reverse genetics, which makes it possible to obtain a mutant phenotype from simple sequence data.

Principle

Forward genetics is based on obtaining mutants for a desired phenotype (e.g., mutant of flower development) in the absence of data on the gene itself or the product for which it codes. When a mutant is created by insertion of the tagging sequence, this sequence can be used by various molecular strategies to recover fragments of the mutant gene that are adjacent to it. These fragments in turn serve as a probe to recover the entire functional gene from a genome library constructed with non-mutant individuals. Forward genetics can thus be used to gain access to the sequence of a gene from a phenotype mutant for this gene. Sophisticated systems have also been developed to target particular types of mutation (see further).

Advances in genome studies have also made it possible to have access, for some model species, to gene sequences for which there are neither phenotype data nor mutants from which the function can be studied. The strategies of reverse genetics developed a few years ago can be used to carry out the gene tagging process backwards, i.e., to access a mutant for a gene from the sequence of that gene. Reverse genetics, like forward genetics, is based on the prior creation of collections of insertion mutants. The mutants in the chosen gene can be located within the collection through the use of a pair of PCR primers, one in the gene sequence and the other in the tag: the amplification between these two primers can take place only if the line carries an insertion in the gene chosen and it will then be possible to make the gene correspond with its associated mutant phenotype. The complexity of the strategy is based on the need to screen a large number of lines: they are clustered in a pyramid of "pools" and "sub-pools" that are analysed sequentially, till individual lines are identified. Reverse genetics is chiefly used for largely sequenced organisms, such as the model plant *Arabidopsis thaliana*, yeast, or *Drosophila*.

Nature of tagging

Two types of tagging can be used: transposons and T-DNA (the latter only in plants). T-DNA[1] is a plasmid DNA fragment originating from the bacterium *Agrobacterium tumefaciens*, which can transfer itself into the genome of an infected plant (see Profile 33). The generation of collections of insertion mutations by T-DNA requires a transformation step, but the insertions are stable.

Transposons are genetic elements that can insert themselves into the genome by means of the functions that are specific to them (at least in the case of autonomous versions). They are differentiated into two classes: (1) elements with DNA, such as Ac elements of maize or P elements of *Drosophila*, which transpose by excising themselves outside the locus of origin, and (2) retroelements, which transpose in a replicating manner through the intermediary of an RNA. The latter have the advantage of creating stable insertions. At present, the most commonly used elements in plants and in *Drosophila* are elements with DNA. The retroelements are essentially used for tagging in yeast and mouse; they are now used also in plants, notably in rice.

In plants, the transposons are used to tag genes, either in species that harbour them (homologous situation) or after they are introduced in the genome of another species (heterologous situation). The heterologous situation is widely favoured, partly because it makes it possible to overcome problems created by the presence of numerous endogenous homologous copies, but also because it allows the use of more sophisticated tags, modified so as to direct the type of mutation or type of function desired (see further).

The chief advantage of transposons lies in their capacity to move on their own, unlike T-DNA, which is inserted only during the transformation step. Moreover, in a heterologous situation, it is possible to control their mobility by using defective versions, which move only when an autonomous version is present at the same time. By clever manipulations of genetic crosses, it is thus possible to make the tagging move and then stop moving at will, and ultimately to make it move again. In this way, it is also possible to overcome the main disadvantage of transposable elements with DNA, the risk of losing the insertion mutation if the element excises itself outside the locus it has tagged.

Identification of genomic sequences flanking the insertion (forward genetics)

The fundamental step of forward genetics is the obtaining of fragments of the gene flanking the tag, which will ultimately serve to clone the entire gene from a genome library constructed with non-mutant individuals. This step has long relied on the construction of genome libraries from the mutant and on screening of the library with a "tag" probe. The positive clones are then analysed to locate and then subclone the partial genome fragment flanking the tag. This type of "standard" cloning is relatively easy when one is working with a heterologous system or with a transposon that is naturally present in a small number of copies (which is rarely the case) and for which one can easily follow the co-segregation of the mutant phenotype and of a fragment identifiable by Southern blot, for example. More recently, PCR strategies such as iPCR or tail-PCR, developed in the hope of using T-DNA, have made it possible to greatly facilitate the cloning of regions flanking the insertions.

In situations in which the transposon is present in a large number of copies, it is more difficult to distinguish the copy that serves as a tag from all the other copies present. A highly sensitive process has been developed thanks to the perfection of refined techniques that allow the simultaneous amplification of a large number of insertion sites and to visualize them together, at high resolution, on an acrylamide gel. These techniques, called transposon display or sequence specific amplified polymorphism (see Profile 53), consist of an amplification of the AFLP type (see Profile 52) anchored in the insertion and make it possible to follow the segregation of a particular DNA fragment within the large number of copies present (several tens or several hundreds).

Modified tagging systems

One of the key conditions for the success of a tagging strategy, whether forward or reverse, consists in a maximal reduction of populations of individuals to be screened before obtaining the insertion in the desired place. Thanks to the use of modified tags, which carry various sequences and functions and allow screening and even selection for a given type of mutation, the chances of success can be significantly improved. It is possible to construct tags that allow easier location of an insertion in a gene or in a promoter, for example, because they carry a reporter gene whose expression will be ensured only after insertion in a coding sequence (gene trap) or in a promoter (enhancer trap). The most widely used reporter genes are colorimetric markers that simultaneously allow study of the mode of expression of the gene tagged. But it is also possible to use selection markers (e.g., resistance to an antibiotic), which allow direct selection of individuals that carry this type of insertion, and not just location of them within the collection. A further refinement is the association of two types of markers within the same tag. It is also possible to find mutations for gain of function, rather than loss, by inserting a strong promoter in the tag that will allow the activation of adjacent sequences (activation tagging). Naturally, these improvements are possible only using modified elements reintroduced in the genome (as well as T-DNA in the case of plants), that is, essentially in a heterologous situation.

[1]Genetic transformation by *Agrobacterium tumefaciens* (carrier of T-DNA) extends to other organisms, notably fungi.

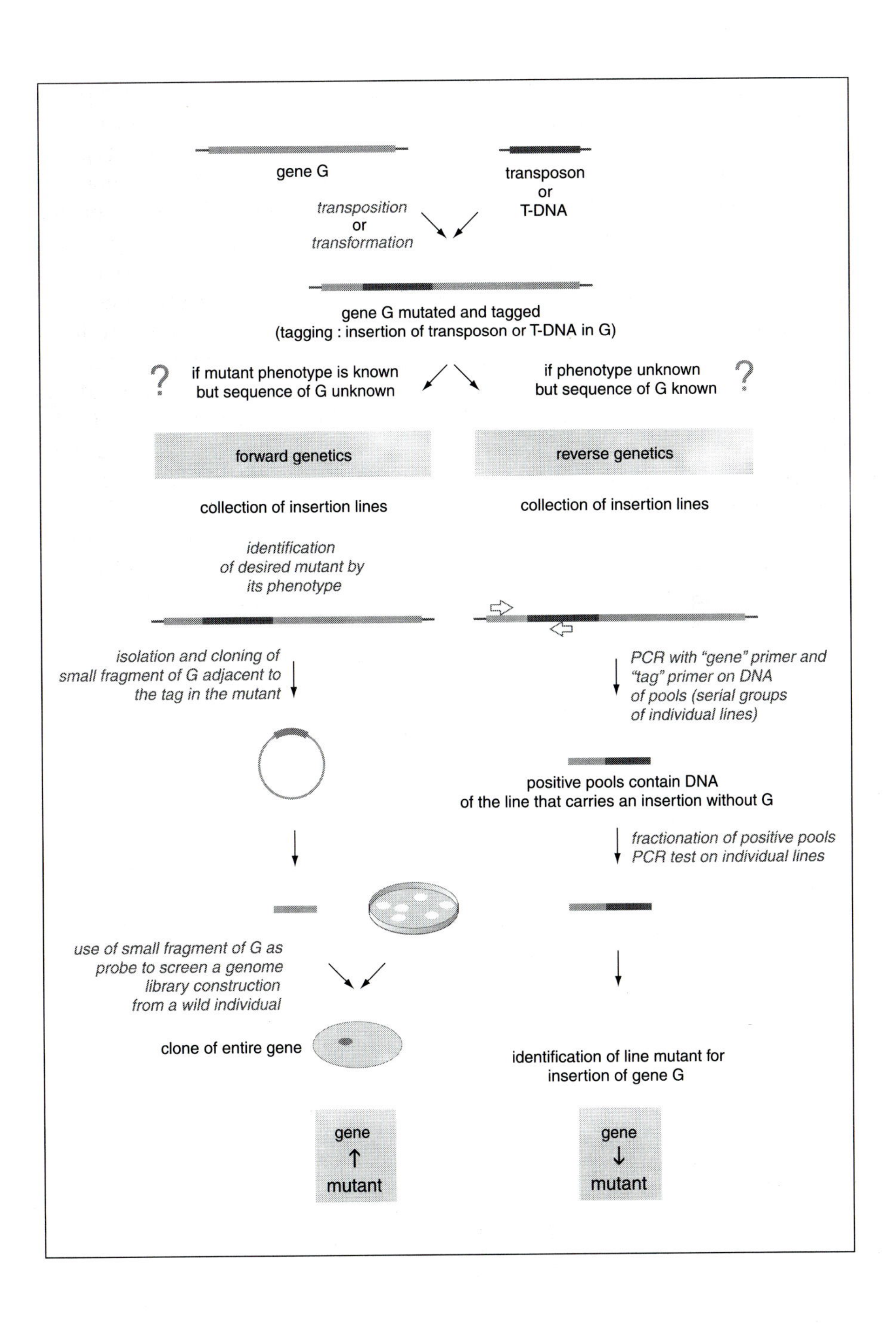
gene G
transposon
or
T-DNA
transposition
or
transformation
gene G mutated and tagged
(tagging : insertion of transposon or T-DNA in G)
?
if mutant phenotype is known
but sequence of G unknown
if phenotype unknown
but sequence of G known
?
forward genetics
reverse genetics
collection of insertion lines
collection of insertion lines
identification
of desired mutant by
its phenotype
isolation and cloning of
small fragment of G adjacent to
the tag in the mutant
PCR with "gene" primer and
"tag" primer on DNA
of pools (serial groups
of individual lines)
positive pools contain DNA
of the line that carries an insertion without G
fractionation of positive pools
PCR test on individual lines
use of small fragment of G as
probe to screen a genome
library construction
from a wild individual
clone of entire gene
identification of line mutant for
insertion of gene G
gene
↑
mutant
gene
↓
mutant

Profile **46**

RNA INTERFERENCE

The use of transgenic plants led to the discovery in the early 1990s of unsuspected (and so far unexplained) mechanisms of silencing of gene expression. In fact, this non-expression was observed during the construction of a transgene that was supposed to overexpress the target gene. How can a desired overexpression result in the silencing of expression of the transgene and the homologous copy of the gene already present in the cell? Or, more trivially, how can 1 + 1 = 0? The artificial input of supplementary copies into the cell seems to block any expression of different copies of this gene. It was subsequently demonstrated that these phenomena occur at the post-transcription level and result from the specific degradation of homologous messenger RNA coming from the endogenous gene and the transgene. The study of these phenomena, called post-transcriptional gene silencing (PTGS), has demonstrated the amplification of double-stranded RNAs (dsRNAs). Their formation results, depending on the case, from the presence of reverse repetitions at the transgenic locus leading to the transcription of sense-antisense mRNAs, or the high production of "sense" transcripts by a mechanism that remains unexplained. Moreover, in 1998, it was demonstrated in the nematode *Caenorhabditis elegans* that the injection of dsRNA induced the inactivation of homologous endogenous genes. This phenomenon, named RNA interference (or RNAi), very quickly showed great similarities with PTGS: (1) intervention of dsRNA in the inactivation of gene expression; (2) similarity between several proteins controlling the phenomena, (3) accumulation of small RNAs of 21 to 25 sense or anti-sense nucleotides homologous to the mRNA of the inactivated gene; and (4) propagation of an inactivation signal that, from a localized initiation, will sometimes induce the inactivation in the entire organism by a systemic response.

The mechanisms at work in phenomena of PTGS and RNAi share many steps. These phenomena also seem to exist in filamentous fungi (such as *Neurospora crassa*), in insects (*Drosophila*) and in mammals. The triggering of

PTGS/RNAi is caused by the presence of dsRNAs of the target gene. These double-stranded molecules are produced from a transgene or directly injected in the cell. They are then cleaved into oligonucleotides of about 20 bp, called small interfering RNAs (siRNAs), by a specific RNase of dsRNA (called DICER in animals). The anti-sense siRNA derived from this cleavage will serve as a guide for an enzymatic complex (RNA-induced silencing complex or RISC), which will degrade the mRNA coming from an endogenous gene. This disappearance of the transcript blocks the translation of the corresponding protein. Moreover, in *C. elegans*, it has been shown that the anti-sense siRNA could serve as primer for an RNA-dependent RNA polymerase (RdRP) for the synthesis of complementary RNA from the mRNA of the endogenous gene, thus enriching the cell in dsRNA and amplifying the RNAi signal.

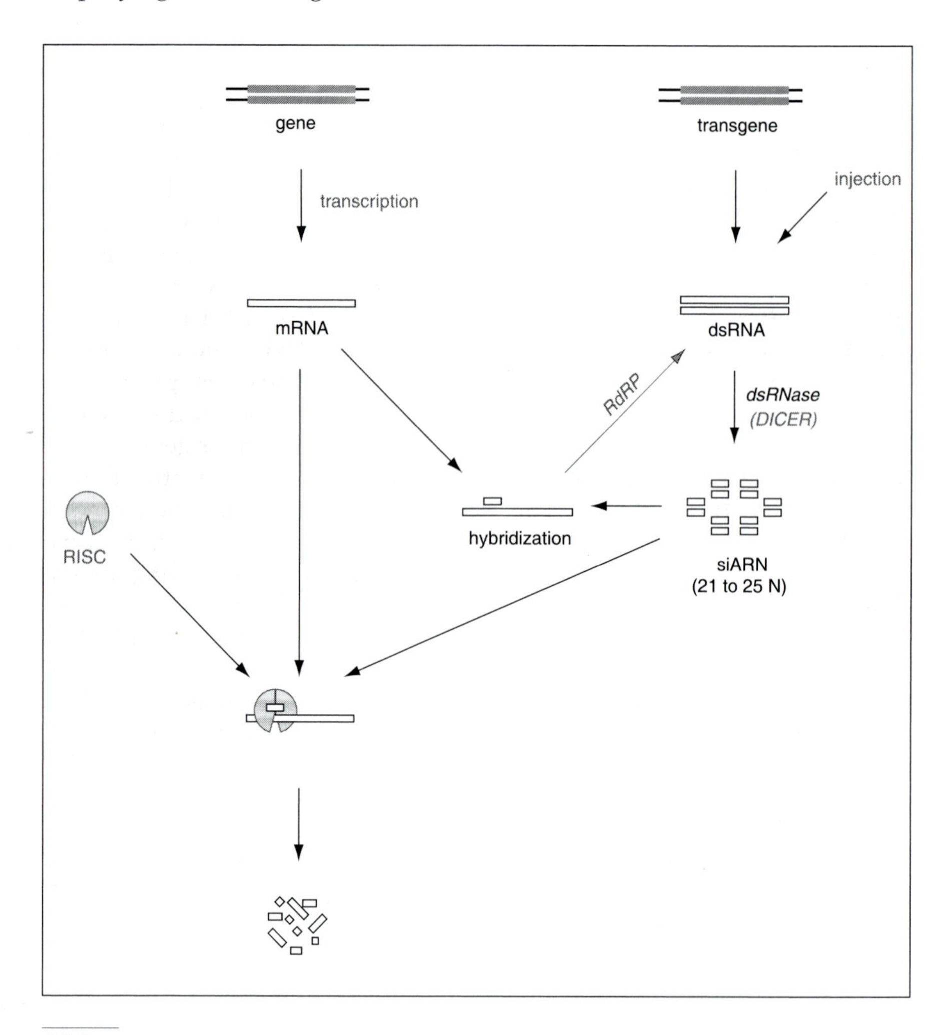

Thus, RNAi can become a relatively easy gene inactivation technique to use, a good alternative to the inactivation of genes by homologous recombination (see Profile 44) that is limited to a very small number of species. The RNAi very quickly was applied to many lower animals. For example, in *C. elegans*, feeding these animals with bacteria containing dsRNAs of target genes was sufficient to block in a transient and near-systemic fashion the expression of this gene in nearly all the cells of the worm. In mammals, the presence of dsRNAs in the cell habitually induces anti-viral responses of the PKR type and/or interferon, leading to the generalized inhibition of translation in the cell. This is why the initial attempts to use RNAi, because of the induction of these non-specific responses after the injection of dsRNA, generally failed. Only the cells or tissues that lacked this type of response (unicellular zygotes, cultures of undifferentiated cells) have responded positively to RNAi. However, this obstacle was successfully overcome by limiting the size of dsRNAs to 21 nucleotides of a clearly identified structure that were capable of causing RNAi without inducing antiviral response.

At present, the use of phenomena of PTGS/RNAi is rapidly increasing. In lower animals (chiefly *C. elegans* and *Drosophila*), the simplicity of the method of injection of dsRNA has made it possible to apply it in genome studies designed to determine simultaneously and gene by gene the function of genes carried by an entire chromosome (around 2500 genes). In mammals, comparable studies are now envisaged on cell cultures.

In plants, transgenes have been produced such that they are transcribed directly into dsRNA. These transgenes are highly effective in inactivating the homologous endogenous genes and producing transgenic plants with a modified phenotype that could have an agronomic use. Moreover, the very high efficiency of the transformation of the model plant *Arabidopsis thaliana* makes it possible to consider the application of these strategies in genome studies.

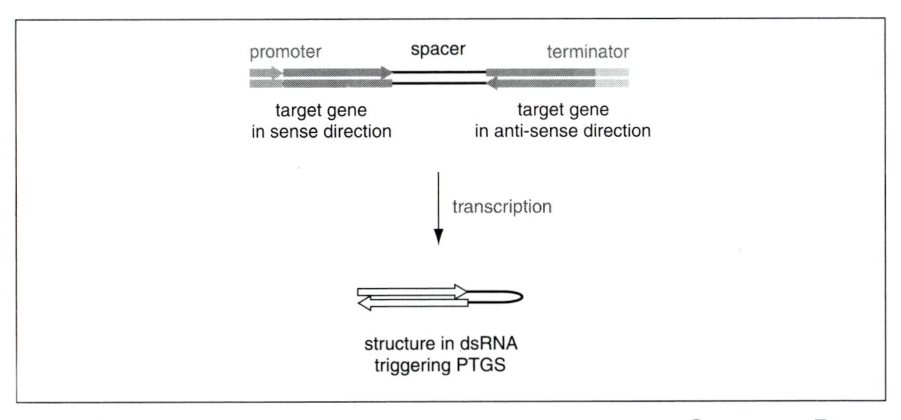

VIII

Polymorphism of a Genome

Profile **47**

MOLECULAR MARKERS

The search for polymorphism of a genome is an approach that has raised high hopes and each year produces new techniques. Each technique has advantages and disadvantages and the choice of technique depends essentially on the nature of the genetic problem to be resolved.

What is a genetic marker?

A genetic marker is a measurable character with Mendelian inheritance. Traditionally morphological markers (e.g., colour, form) are distinguished from molecular markers (at the level of DNA) and biochemical markers (isozymes, proteins, secondary metabolites such as terpenes). The characteristics of an ideal marker can be summarized as follows:

- Polymorphic: variable between individuals.
- Discriminating: allowing very similar individuals to be differentiated.
- Multiallelic: having several alleles (thus different DNA sequences) on the same locus.
- Codominant: a heterozygous individual presents simultaneously the characters of two homozygous parents and it can thus be distinguished from each of the parents.
- Non-epistatic: the genotype of the marker can be determined from its phenotype, no matter what the genotype at other loci may be. In other words, the reading of the marker studied is independent of the expression of other markers of the genome.
- Independent of the environment: the genotype of the marker can be determined from its phenotype in any environment.
- Neutral: no matter what the allele present at the marker locus, the selective value of the individual remains the same.

- Distributed uniformly in the entire genome: in order to evaluate the polymorphisms in multiple places in the genome.
- Reproducible from one experiment to another.
- Economical and capable of being manipulated on a large scale.

Morphological markers are mostly poorly polymorphic and generally dominant. They can be influenced by the environment and interfere with other characters. The limitation of biochemical markers is the low number of loci revealed as well as a certain specificity of the organ and/or stage of development. On the other hand, markers at the level of DNA are nearly unlimited in number and are independent of the stage of development or organ analysed. Some typing techniques reveal molecular markers "individually": one reaction produces a single molecular marker (e.g., microsatellites). Other techniques reveal markers en masse: a reaction thus provides several markers (e.g., RAPD or AFLP, see Profiles 51 and 52).

Principal sources of molecular markers

Sequence polymorphism

Sequencing is the only exhaustive method of searching for polymorphism in a DNA fragment. Despite advances in automation, the routine use of this technique remains rare, because of the number of loci and the large numbers generally needed in genome studies. Indirect methods are therefore used, which are not exhaustive but are much faster and more economical.

- Codominant markers revealed individually; the techniques are based on detection of the following:
  - differences at the level of restriction enzyme sites: RFLP (see Profile 50) and CAPS;
  - differences of DNA conformation: SSCP (see Profile 54);
  - differences of DNA stability: DGGE/TGGE technique (see Profile 55).
- Dominant markers revealed en masse: the techniques are based on detection of the following:
  - differences of hybridization site from one arbitrary primer (a mutation on the binding site of a primer prevents its binding and results in the disappearance of a band): MAPP (RAPD, AP-PCR, DAF);
  - differences of cleavage sites and hybridization sites of an arbitrary primer: AFLP and tecMAAP.

Polymorphism of number of repeats

- Microsatellites

Microsatellites or simple sequence repeats (SSR) are made of tandem repeats of mono-, di-, tri-, or tetranucleotide motifs, for example (A)n, (TC)n,

(TAT)n, (GATA)n, the value of n varying from a few units to several tens. Microsatellites are characterized by extremely high polymorphism. The polymorphism involves several repeat units. The PCR technique is used to locate the microsatellite loci that are codominant and multiallelic markers. Although a microsatellite is not specific to a locus, the flanking regions are. A pair of primers specific to these flanking regions will thus amplify this single microsatellite, and thus the length polymorphism is revealed in a high resolution acrylamide gel (see Profile 57).

- Inter-microsatellite amplification (IMA)

A more recent exploitation of microsatellites consists of revealing them en masse, following the principle of RAPD. For this, a primer made up of a microsatellite sequence and arbitrary bases is used.

- Minisatellites

Minisatellites, which have the same overall structure as microsatellites, are distinguished by the length of the repeat unit, which can be several tens of bases.

Choosing the right marker

One of the applications of genetic markers is the establishment of genetic maps (see Profile 48) that allow the location of chromosomal regions that control a quantitative trait locus (QTL). The use of these data in a genetic improvement programme constitutes one of the facets of marker-assisted selection (MAS).

- The dominant markers MAAP (multiple arbitrary amplicon profiling) and AFLP can be used to generate markers en masse and are well suited to objectives that require many markers. They are thus used for the following purposes:
  - to rapidly saturate a genetic map in species that have not been thoroughly studied in which RFLP probes or STS and other codominant markers have not been developed;
  - in population genetics, for studies in diversity: e.g., intra- or interpopulation structuration, genetic distances, classifications;
  - for positional cloning, where they can be used to obtain a high marker density near the gene to be cloned (chromosome walking).
- Codominant makers, specific to a locus, are also widely used.
  - codominance allows the dominance at QTL to be measured;
  - if the markers are multiallelic, they will provide information on many pedigrees and can be widely used for construction of genetic maps;
  - if they can be used in phylogenetically close species or genera, they can be useful in comparative mapping.

markers	*neutrality*	*number*	*co-dominance*	*locus specificity*[1]	*poly-morphism*	*stage, organ*[2]	*technicity*	*cost*	*coding sequence*
morphological	no	limited	rare	yes	low	yes	low	low	?
isozymes	yes	< 40	yes	yes	low	yes	low	low	yes
proteins (2D)	yes	< 100	yes	yes	low	yes	medium	medium	yes
RFLP	yes	unlimited	yes	yes	high	no	high	medium	yes/no
MAAP	yes	unlimited	no	no	very high	no	low	low	yes/no
AFLP	yes	unlimited	no	no	very high	no	medium	medium	yes/no
microsatellites	yes	few	yes	yes	very high	no	high	very high[3]	yes/no
IMA	yes	unlimited	yes	no	very high	no	low	low	no
STS	yes	few	yes	yes	medium	no	high	medium	yes

[1] In this column, techniques that reveal the marker loci individually are distinguished from those that reveal them en masse.
[2] Markers linked to the stage of development or to the organ studied.
[3] Mostly for those steps necessary for the development of microsatellites.

		Description	*Reference*
AFLP	Amplification fragment length polymorphism	PCR amplification of genomic DNA after cleavage by two restriction enzymes and ligation of an adaptor of around 20 bp. The primers are made up of the adapter and three random bases on the 3' end	Vos et al. (1995), *Nucl. Acids Res.*, 23: 4407-4414 (see Profile 52)
AP-PCR	Arbitrary primed PCR	PCR amplification of unknown sequences using a single primer of 20 arbitrary bases, with low hybridization temperatures	Welsh and McClelland (1990), *Nucl. Acids Res.*, 19: 861-866
CAPS	Cleaved amplified polymorphic sequence	An amplified fragment of genomic DNA (STS or SCAR) is cleaved by restriction enzymes, then the fragment length polymorphism generated is revealed by electrophoresis	Konieczny and Ausubel (1993), *Plant J.*, 4: 403-410
DAF	DNA amplification fingerprinting	RAPD using shorter primers (5 to 10 bases)	Caetano-Anolles et al. (1991), *Bio/technology*, 9: 553-557
DGGE	Denaturing gradient gel electrophoresis	Electrophoresis on an acrylamide gel containing a linear gradient of urea/formamide. The fragment of DNA, initially double stranded, is partly denatured to form a "branched" or Y-shaped structure that is nearly immobile. Some differences in sequence in the least stable fusion domain lead to differences in stability and thus level of migration	Myers et al. (1987), *Methods Enzymol.*, 155: 501-527
IMA	Inter-microsatellite amplification	PCR amplification on genomic DNA with primers that carry dinucleotide repeat motifs and some arbitrary bases	Zietkiewicz et al. (1994), *Genomics*, 20: 176-183 (see Profile 57)
MAAP	Multiple arbitrary amplicon profiling	Generic term designating all the PCR amplification techniques on genomic DNA using arbitrary primers generating complex electrophoretic and polymorphic profiles (RAPD, AP-PCR, DAF, AFLP, IMA...)	Caetano-Anolles et al. (1994), *Plant Mol. Biol.*, 25: 1011-1026

Table contd.

Table contd.

tecMAAP	Template endonuclease cleavage MAAP	PCR amplification of the MAAP type on genomic DNA that is previously cleaved by a restriction enzyme. The AFLP technique belongs to this category	Caetano-Anolles et al. (1993), *Mol. Gen. Genet.*, 241: 57-64
Microsatellite or SSR	Simple sequence repeat	Tandem repeats of mono-, di-, tri-, or tetranucleotide motifs at different loci. Such motifs are abundant and highly polymorphic in the genome of eukaryotes	Tautz (1989), *Nucl. Acids Res.*, 17: 6463-6471 (see Profile 57)
RAPD	Random amplified polymorphic DNA	PCR amplification of DNA fragments using a primer of 10 arbitrary bases	Williams et al. (1990), *Nucl. Acids Res.*, 18: 6531-6535 (see Profile 51)
RFLP	Restriction fragment length polymorphism	Hydrolysis of DNA by a restriction endonuclease then polymorphism revealed after hybridization of a marked probe on fragments separated by gel electrophoresis and transferred on a membrane (Southern technique)	Botstein et al. (1980), *Am. J. Human Genet.*, 32: 314-331 (see Profile 50)
SCAR	Sequence-characterized amplified region	Genomic DNA fragment amplified by PCR with specific primers (14 to 20 bases). These primers are defined according to the knowledge of the sequence of an interesting RAPD fragment isolated on a gel, cloned, and sequenced	Paran and Michelmore (1993), *Theor. Appl. Genet.*, 85: 985-993 (see Profile 51)
SSCP	Single-strand conformation polymorphism	Genomic DNA fragment amplified by PCR with specific primers. The double-stranded DNA is denatured (94°C, 10 min.), rapidly cooled in order to prevent the rejoining of single strands, and charged on a non-denaturing acrylamide gel	Orita et al. (1989), *Proc. Natl. Acad. Sci. USA*, 86: 2766-2770 (see Profile 54)
STS	Sequence tagged site	PCR amplification with specific primers of a sequence	Olson et al. (1989), *Science*, 254: 1434-1435

Profile **48**

GENETIC AND PHYSICAL MAPS

A genetic map is a simplified representation of the genome that consists of lining up all the markers of chromosomes and determining the distances between them. The unit of measurement used (the centimorgan, cM, or unit of genetic distance) is calculated from the frequency of recombination between loci (genetic markers) taken two at a time. In the case of a physical map, the DNA molecule is marked out from molecular markers made up of particular sequences and the distances are expressed in number of base pairs (bp).

Establishment of genetic map

In practice, the establishment of a genetic map is based on the phenomenon of crossing over at meiosis. The creation of a genetic map requires (1) a progeny line from a cross obtained by sexual reproduction and (2) determination of the genotype of different markers for individuals of that line. Statistical analysis and methods of mathematical calculations can then be used to estimate the genetic distances between markers.

The greatest utility of molecular markers has been in the development of genetic linkage maps. Saturated genetic maps make possible detailed genetic studies, decomposition of complex characters into their discrete components (quantitative trait loci or QTL), and the use of a linkage between marker and QTL for marker-assisted selection. Since the late 1980s, RFLP maps (see Profile 50) of major agronomic species have been established. During the past ten years, technological advances in the detection of molecular markers (mainly the advent of PCR-derived markers) as well as the development of statistical analysis of data from markers have led to a virtual explosion in the field of genetic mapping. Existing maps have been completed (the number of linkage groups is equal to the number of chromosomes). Most of all, these maps make it possible to study species on which no genetic information existed or for which no classical pedigree could be used to establish a map (e.g., species having a long gestation period, such as forest trees).

Establishment of a physical map

The establishment of a physical map involves reconstructing a genome in an ordered set of DNA fragments as they are found on chromosomes. The cloning vectors used are of the cosmid type, BAC or YAC (see Profile 8), since they allow the integration of large DNA fragments. The physical map is often an indispensable prerequisite to genome sequencing. In practice, a physical map is constructed through the tagging of each clone by molecular markers (DNA sequences), if possible the same ones used to obtain the genetic map. The difficulty lies in organizing the clones among themselves in order to discover their initial agency on the chromosome. The clones with common parts, revealed by common markers, are clustered into "contigs", i.e., into sets of overlapping clones. All the "contigs" together then allow the reconstruction of the chain of DNA fragments constituting the chromosomes.

Relationship between genetic and physical maps

The integration of the physical and the genetic map of a given genome consists in placing numerous points of location that are common to the two maps: for example, RFLP markers or microsatellites located on a genetic map can be easily sequenced and located on a physical map. The difficulty of integration is linked to the fact that the two scales used (cM on the one hand and number of base pairs on the other) are not colinear. In *Arabidopsis*, 1 cM corresponds to 220 kb on average but varies with the regions of the genome from 30 to over 1000 kb. Moreover, even if the order of markers is respected between the two types of map, their relative distances are sometimes very different. This is due to the fact that the frequency of recombination is not constant over the entire genome. There are regions that recombine often (recombination hot spots), while the physical distances between them are small. On the other hand, there are very long portions of DNA (centromeric and telomeric regions) in which the phenomenon of recombination is suppressed by the presence of heterochromatin or highly repeated DNA. These zones consequently show smaller genetic distances between the markers, which tend to group together into clusters. In other words, in a poorly recombined region of the genome, 1 cM corresponds to a large number of bases, while in a highly recombined region 1 cM corresponds to fewer bases.

The ratio of physical distance to genetic distance varies at the intra-chromosomal level, but it varies more generally when species with different DNA levels are compared. The table that follows presents the relationship between physical and genetic distance in some species for which complete genetic maps are available. It shows the following:

- The genetic size of a genome varies little (from single to double) and comes to about 150 cM (= 1.5 Morgan) per chromosome. Thus, meiosis is represented as a process that fractures the genome slightly with a mean of 1.5 crossing-over per chromosome.
- The low variation of lengths of genetic maps between species contrasts with the considerable variations of quantity of DNA per cell. As the number of genes expressed must be of the same order of magnitude within higher plants, these differences in the quantity of DNA will essentially be due to very high variation in the quantity of repeat DNA that is not transcribed. These variations in the physical size of genomes seem independent of genetic sizes. Thus, it appears that the frequency of crossing-over per unit of physical length decreases when the size of the genome increases.

A comparison between genetic and physical maps of a given genome makes the following possible:

- From the position of an interesting locus on the genetic map, easier access is gained to the corresponding genes (positional cloning).
- From that same position, the genetic position of any DNA sequence can be obtained.

Physical mapping is costly and is presently considered only for model species with a "small genome" (*Arabidopsis*, rice) for which genetic maps are also available. However, thanks to comparative mapping studies, some of the knowledge obtained in model species can be transferred to species of agronomic interest, for example, from *Arabidopsis* to rape, or from rice to wheat.

Species	*Physical size (Mb)*	*Genetic size (cM)*	*No. of chromosomes (n)*	*Avg. genetic size per chromosome (cM)*	*Mb/cM*
Phage T4	0.16	800	/	/	0.0002
Escherichia coli	4.6	1750	/	/	0.0026
Saccharomyces cerevisiae	14	4200	16	262.5	0.0033
Caenorhabditis elegans	100	320	11/12	/	0.312
Arabidopsis thaliana	150	675	5	135	0.22

Table contd.

Table contd.

Drosophila melanogaster	170	280 F	4	70	0.607
Prunus persica	300	712	8	90	0.42
Oryza sativa	450	1490	12	125	0.30
Eucalyptus grandis	600	1370	11	125	0.43
Brassica rapa	650	1850	10	185	0.35
Quercus robur	900	1200	12	100	0.75
Lycopersicon esculentum	980	1280	12	107	0.76
Solanum tuberosum	1540	1120	12	93	1.37
Zea mays	2500	1860	10	186	1.34
Lactuca sativa	2730	1950	9	217	1.40
Homo sapiens	3000	2800 M 4800 F	23	121.7 M 208.7 F	1.071 M 0.625 F
Mus musculus	3000	1700	20	100	1.764
Triticum tauschii	4200	1330	7	190	3.15
Hordeum vulgare	5500	1250	7	178	4.40
Triticum aestivum	16,000	3500	21	167	4.57
Pinus pinaster	25,500	2000	12	167	12.75

M, males; F, females.

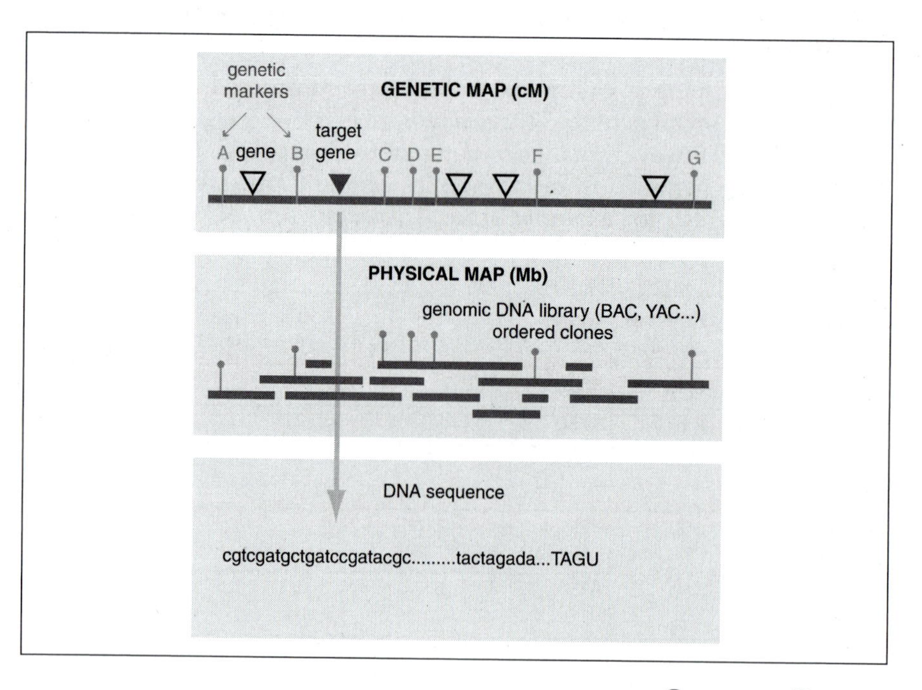

Profile 49

PFGE: PULSE FIELD GEL ELECTROPHORESIS

Conventional agarose gel electrophoresis (with a constant electric field) allows the separation of DNA fragments of size varying from a few hundreds of nucleotides to 50 kb. Beyond that size, the fragments are no longer separated and form a migration front.

Pulse field gel electrophoresis (PFGE) was developed by Carle and Olson in 1985 for the study of yeast *Saccharomyces cerevisiae* chromosomes. The technique makes it possible to separate DNA fragments of size varying from a few kilobases to over 10,000 kb. The most frequently used size standards are either multimers of the genome of the lambda bacteriophage (from 50 to more than 1000 kb) or chromosomes of *Saccharomyces cerevisiae* (between 250 and 2500 kb). The DNA fragments are subjected to alternate electric fields in which the relative orientation may vary from 90° to 180°, depending on the systems. The most widely used configuration at present is contour-clamped homogeneous electric fields (CHEF), in which the angle between two electric fields is 120°. The principle of separation is based on the differential speed of reorientation of molecules at the time of alternation of two electric fields. Thus, the large molecules will take more time than small molecules to orient themselves according to the direction of the second electric field. The accumulation of delay at each alternation leads to the separation of the molecules at the end of the migration. Parameters such as agarose percentage, tension applied, or the temperature can be varied to optimize the resolution. However, the principal parameter is the pulsation time, that is, the time of application of an electric field in each direction: this favours the separation of fragments in a given size range.

PFGE requires a specific process of DNA purification. To realize a karyotype (set of chromosomes of an organism) or a profile of macrorestriction of a genomic DNA (DNA fragments generated by a restriction endonuclease at low cleavage frequency and allowing the visualization of the entire genome) requires the preparation of a DNA molecule of high molecular weight. The preparation of samples prohibits conventional purifications by phenol or chloroform extraction, which favours random breakages of the DNA. This is why the cells are packaged in

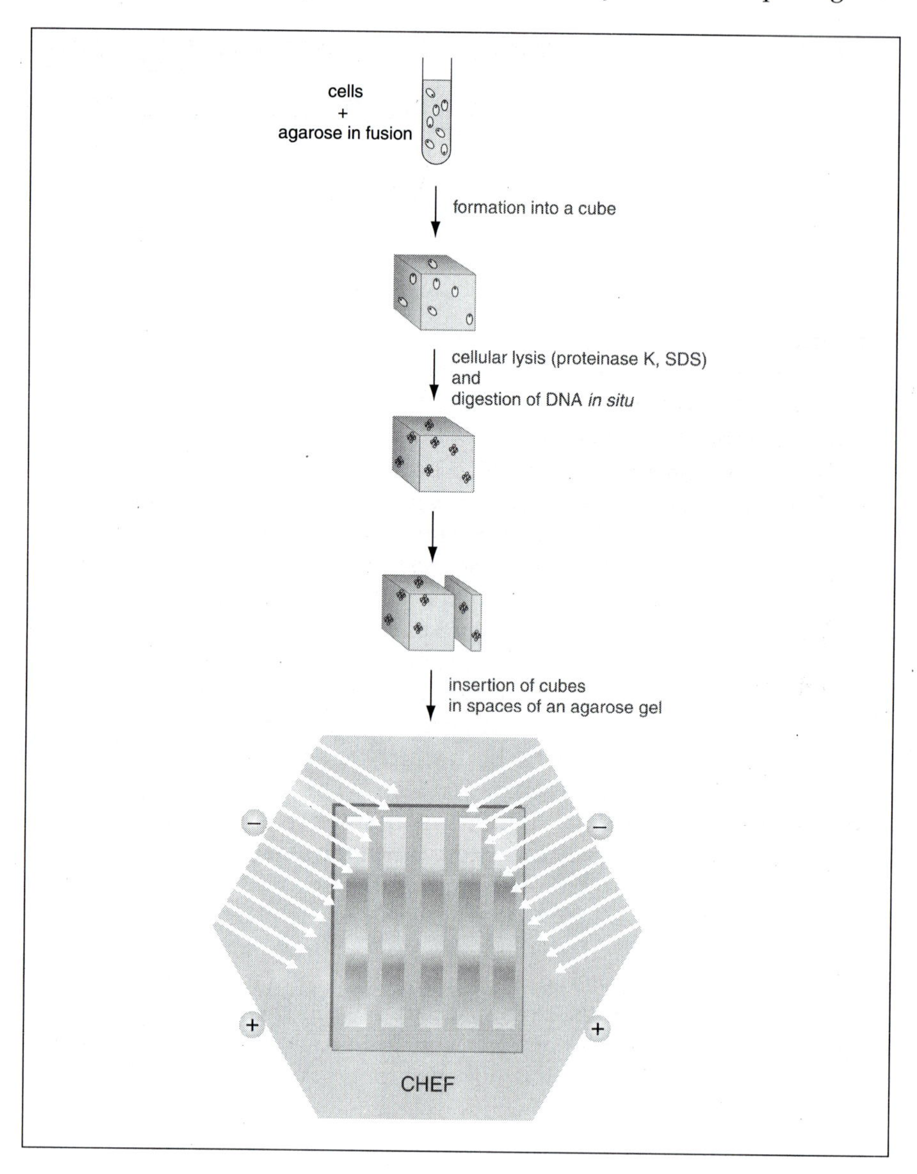

a matrix of agarose at a low fusion point and then lysed *in situ* (by lysozyme for bacteria or other enzymes such as lyticase for yeasts) in order to degrade the outer cell coat. Then the joint action of a detergent and a proteinase releases the cell contents and degrades the cell proteins (including proteins joining the DNA), which will diffuse towards the outside of the agarose matrix. The DNA remains trapped in the agarose. In the same way, cleavage by restriction enzymes (chosen as a function of their cleavage frequency for the DNA studied) will be done within an agarose block protecting the DNA.

The technique has many applications: identification and monitoring of microorganisms (applications in epidemiology, agro-foods), mapping of human chromosome fragments, study of the structure and evolution of bacterial chromosomes. In the last case, the construction of physical chromosomal maps (more than 100 maps since the development of PFGE) has opened up avenues for the systematic sequencing of genomes.

Profile 50

RFLP: RESTRICTION FRAGMENT LENGTH POLYMORPHISM

The polymorphism of DNA is used as a genetic marker for identification of individuals in numerous domains. Within a population, polymorphism, defined as the result of variations in the DNA sequence, is expressed either as the appearance of different phenotypes or by modifications of restriction profiles. Within the genome, different phenomena that are more or less complex are responsible for polymorphism:

- Mutations due to replacement of one or several bases by others without change in the length of the DNA.
- Insertions or deletions corresponding respectively to the addition or disappearance of bases in the DNA sequence.
- Chromosomal rearrangements due to genetic recombination or the insertion of transposons. Genetic recombination is the consequence of a breakage of the DNA followed by an exchange of material between two chromosomes. This phenomenon is called *crossing-over* when the material is exchanged during meiosis. Transposons are DNA sequences capable of replicating themselves and inserting one of their copies into a new site on the genome.

The principle of RFLP lies in the comparison of profiles of cleavage by restriction enzymes following the existence of a polymorphism in the sequence of a DNA molecule in relation to another (e.g., the DNA of a father and his son). Mutations appearing on a DNA sequence recognized by a restriction enzyme cause different lengths of restriction fragments. The DNA of individuals to be compared is thus cleaved by one or several restriction enzymes. The products of restriction are then separated on an acrylamide or agarose gel in the presence of a molecular weight marker. The RFLP observed is used as a criterion of identification. Through this approach, two individuals can present different restriction profiles.

Numerous techniques are used to detect a polymorphism:

- Southern molecular hybridization: the restriction fragments are hybridized with a probe constituted of a marked fragment of DNA complementary to a specific DNA sequence (see Profile 12).
- PCR: a specific region of the DNA is amplified, which allows visualization of restriction fragments of this region, present in large quantity, after gel electrophoresis and staining with ethidium bromide (see Profile 24).
- RAPD: fragments are amplified at random and detected after agarose gel electrophoresis and staining with ethidium bromide (see Profile 51).
- Any other PCR technique using especially repeat sequences (transposons, microsatellites: see Profiles 49 to 57).

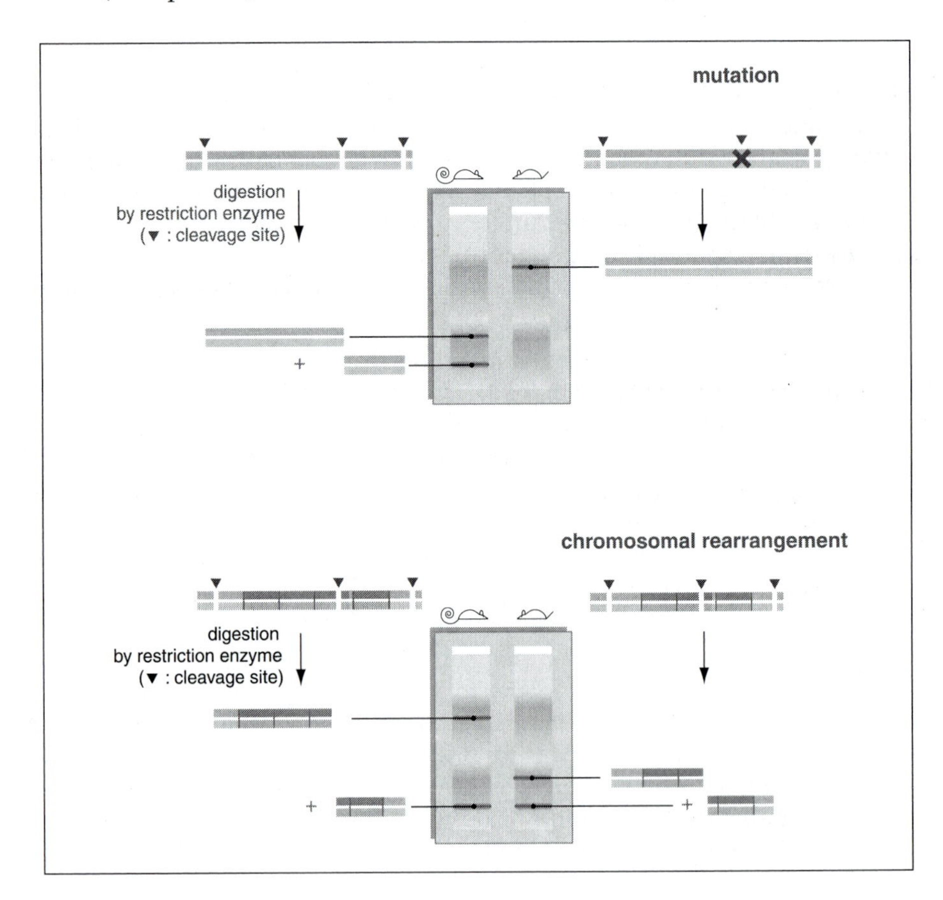

Profile 51

RAPD: RANDOM AMPLIFIED POLYMORPHIC DNA

During a search for polymorphism in a genome, if no precise DNA sequence is targeted, small oligonucleotides (6 to 9 bases) of random sequence can be used that will fix on the target DNA at random. PCR (Profile 24) can be used to obtain products of amplification of an unknown sequence.

This technique is called random amplified polymorphic DNA (RAPD). Unlike with conventional PCR targeted on an identified sequence, a RAPD reaction uses a single type of primer and not a pair of different primers. If two neighbouring primers are in correct position, in the proper orientation and not too far from one another (less than 5000 bp), the PCR generates a product of amplification. RAPD can be used to quickly obtain various fragments of genomic DNA that are then analysed. The conditions of sample preparation and PCR cycles must be optimized for each application in order to ensure the reproducibility of the results.

The amplification fragments permitting the location of a polymorphism can be sub-cloned (see Profile 9) in order to be sequenced. From this sequence, it is possible to redefine a pair of primers that will have a high probability of being specific to the sample analysed. This is a sequence characterized amplified region or SCAR.

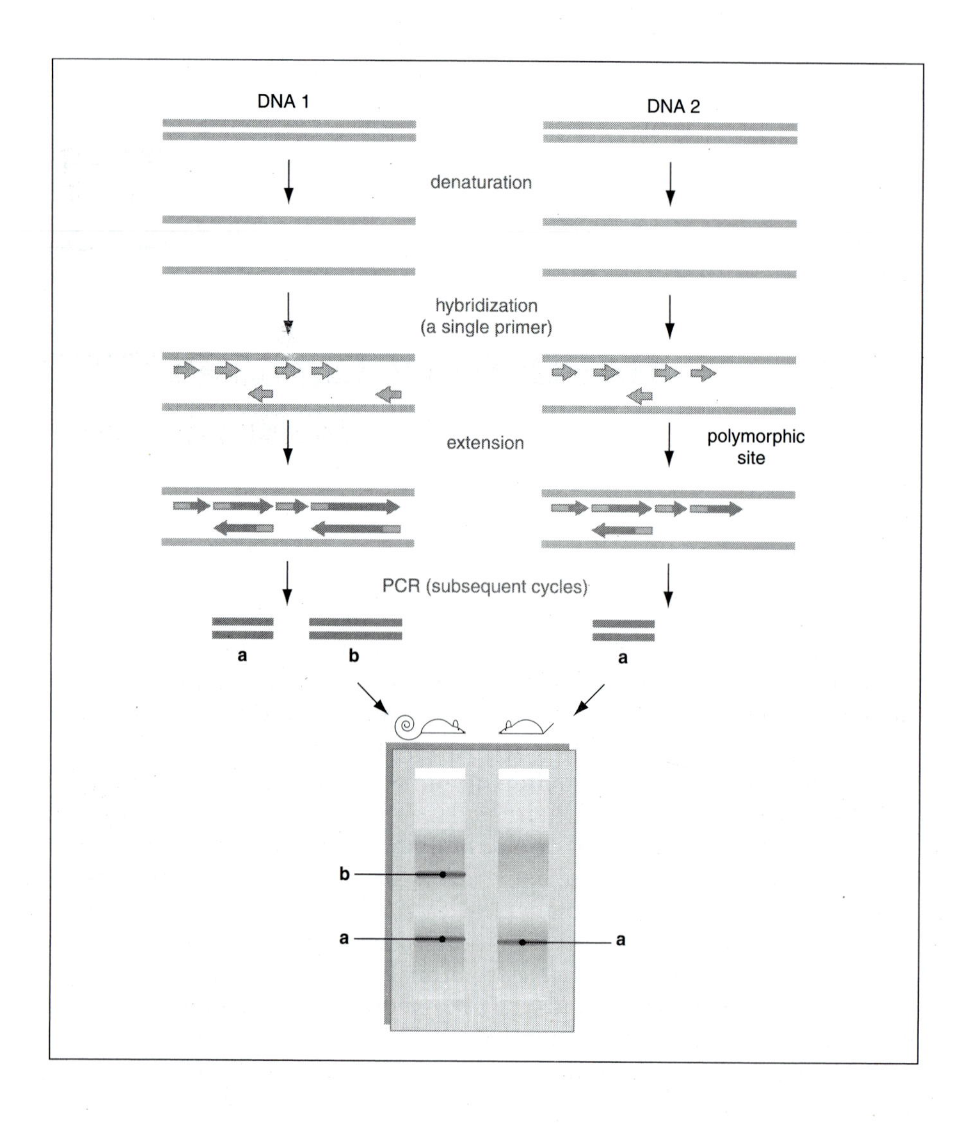
DNA 1
DNA 2
denaturation
hybridization
(a single primer)
extension
polymorphic
site
PCR (subsequent cycles)
a
b
a
b
a
a

Profile 52

AFLP: AMPLIFIED FRAGMENT LENGTH POLYMORPHISM

The technique of amplified fragment length polymorphism (AFLP) was introduced in 1993. Its principle is the selective amplification of restriction fragments generated from a sample of genomic DNA. This technique makes it possible to look for restriction fragment length polymorphism at the level of DNA. It is used for example to identify species, analyse pedigree, or search for genetic markers linked to a character. It is also used to identify genes expressed in a differential manner. In this case, the AFLP analyses are carried out on complementary DNA (cDNA) synthesized from messenger RNA (mRNA). This technique, called AFLP-cDNA, makes it possible to visualize sub-populations of mRNA indirectly and compare them with one another (see Profile 17).

Genomic DNA (or cDNA) are cleaved by two restriction enzymes that respectively cleave at a rare site and a frequent site. Adaptors specific to the cleavage sites used are then fixed at the ends of the restriction fragments obtained. The amplifications are carried out using primers defined according to the sequence of adaptors. These primers carry on their 3'-OH ends random extensions of 1 to 3 bases allowing selective amplification of only part of the population of restriction fragments. A pre-amplification is carried out using primers presenting an extension of generally one base (+1 primers). The product of this pre-amplification serves as a matrix for the selective amplification using +3 primers. For a single pair of enzymes, the use of +3 primers has 4096 (64×64) possible combinations. The products of selective amplification are then separated on a polyacrylamide gel under denaturing conditions. The detection of AFLP bands is obtained either by silver staining of gels or by autoradiography or fluorescence emission. In this case, one of the +3 primers of AFLP (generally that corresponding to the rare cleavage site) is labelled either radioactively (γ-^{32}P or γ-^{33}P) or by a fluorescent chromophore (in the case of analysis using automated sequencer).

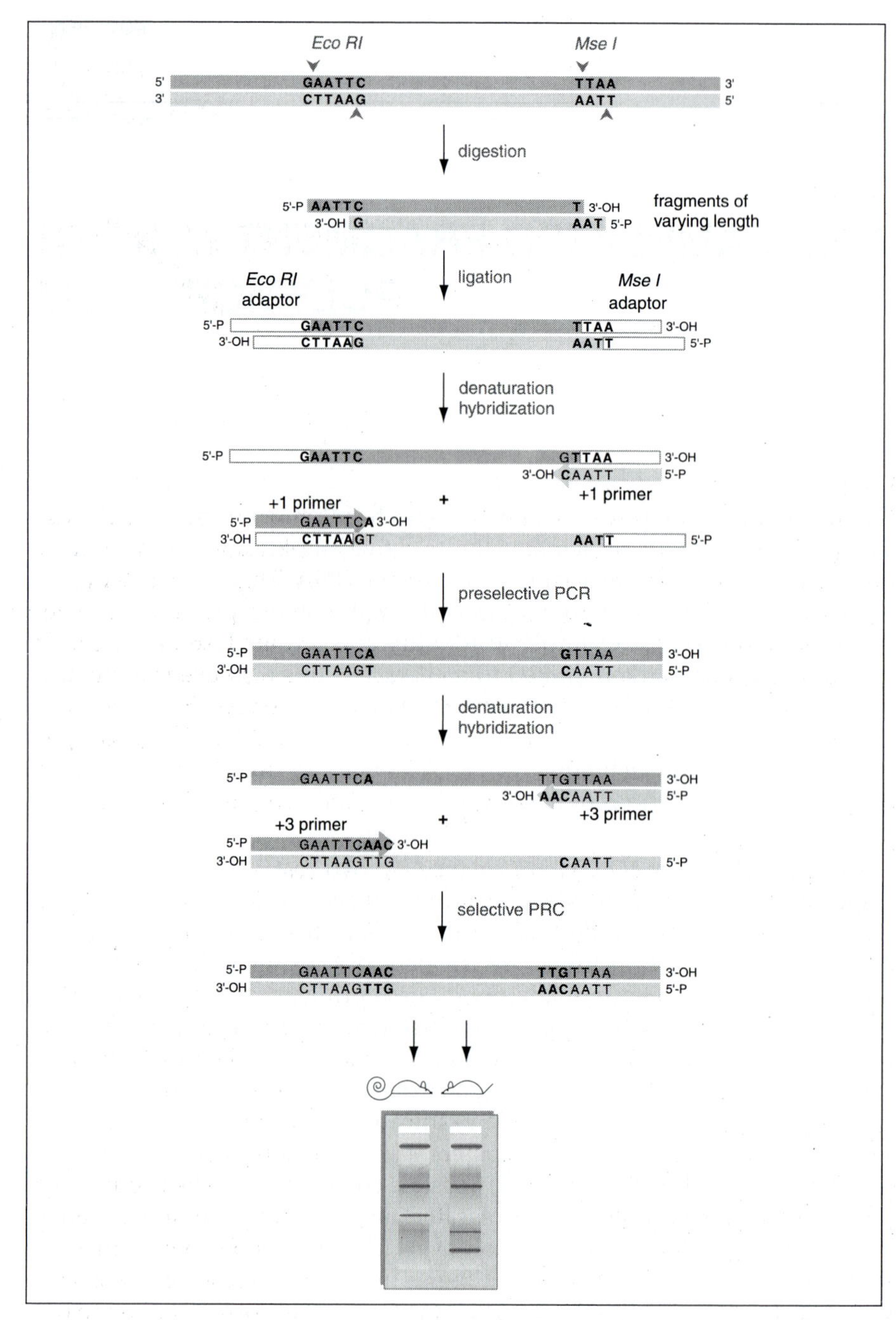

Profile **53**

RETROMARKERS

Retroelements are a major source of genetic variability in eukaryotes. It is a particular class of repeat mobile DNA that is amplified via reverse transcription of an intermediary RNA; it combines various types of elements, from small retroposons of the SINE type (short interspersed nuclear elements) to long retroposons with LTR (long terminal repeat). They are generally present in a large number of copies (several hundreds or even thousands) and represent more than 70% of the genome in some species. Because of their replicative mode of transposition, retroelements create

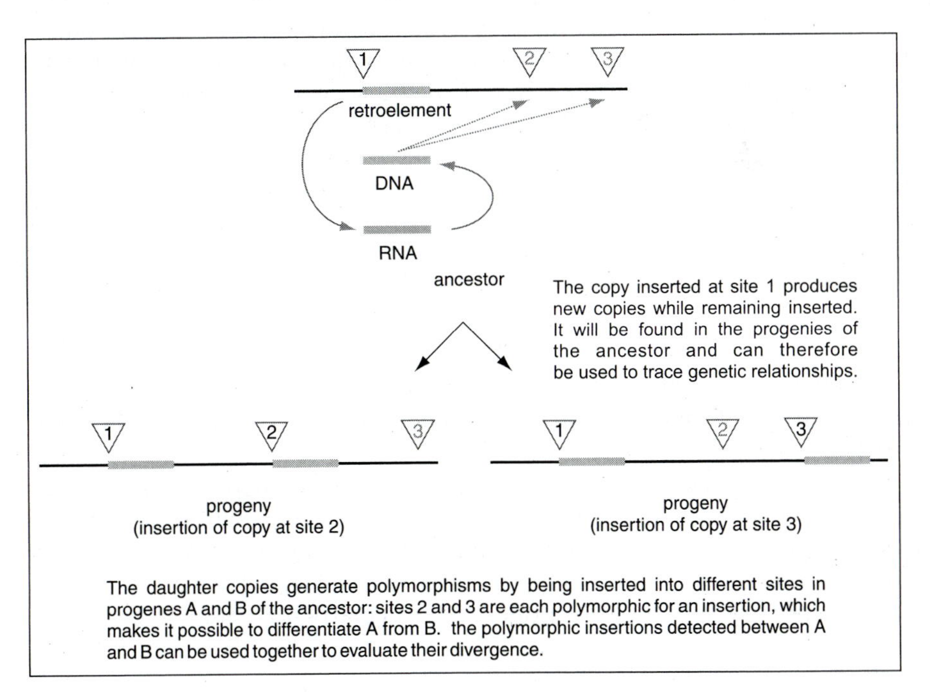

The daughter copies generate polymorphisms by being inserted into different sites in progenes A and B of the ancestor: sites 2 and 3 are each polymorphic for an insertion, which makes it possible to differentiate A from B. the polymorphic insertions detected between A and B can be used together to evaluate their divergence.

irreversible insertions, making it possible to trace genetic relationships. On the other hand, their insertion polymorphisms, which reflect events of transposition that have taken place after the divergence between two genotypes, can be used to evaluate divergences between these genotypes.

Depending on the moment at which new insertions have taken place, these polymorphisms can serve to identify either a given line or a group of lines, or even a species or group of species. Besides, the polymorphisms generated by the insertion of a retroelement correspond to unidirectional events (a state that was earlier "empty" but is now "full", the return towards the ancient state by excision being theoretically impossible), which makes them useful tools of molecular systematics, widely used for a long time in animal systems and more recently in plants.

Practical strategies

Multiplex strategies

Multiplex strategies are based on the use of PCR and generate several bands of amplification per sample. They make it possible to detect all (or at least a significant part) of the insertion sites of a given element. The most widely used is *sequence specific amplified polymorphism* (SSAP), also called *transposon display* (TD). SSAP is a strategy derived from AFLP (see Profile 52) but using, apart from the "adaptor" primer, a labelled primer (radioactivity, fluorescence) corresponding to a terminal sequence of the retroelement, so as to visualize only the DNA fragments beginning in the retroelement. The principle is based on the extraction of genomic DNA followed by its enzymatic digestion by a restriction enzyme. An adaptor is fixed at the ends of DNA molecules. Then a PCR is carried out using a primer complementary to the adaptor and a primer complementary to one of the ends of a retroelement. This supposes therefore that a retroelement has already been characterized and sequenced in the species studied. The DNA fragments resulting from this amplification are analysed for example by electrophoresis on acrylamide gel with a high resolution in order to detect a size polymorphism of fragments of amplification.

SSAP can be used to define a map of the state of representation of the retroelement in the genotype studied and to detect polymorphisms of insertion through the presence or absence of particular SSAP fragments. It thus serves as a tool to evaluate biodiversity and establish phylogeny: it has proved more informative than conventional molecular markers in several cases.

Other multiplex strategies, such as inter-retrotransposon amplified polymorphism (IRAP) or retrotransposon-microsatellite amplified polymorphism (REMAP) are based on direct PCR between retroelements or between retroelements and microsatellites respectively. They are easier to implement than SSAP but presuppose a particularly high number of copies

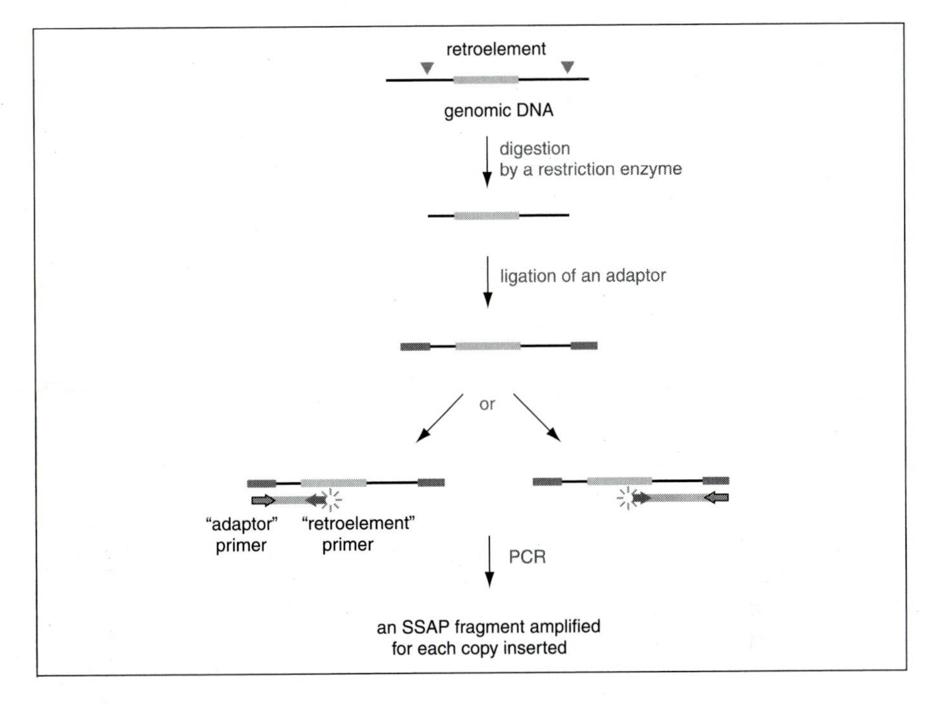

of the retroelement (several tens of thousands of copies, which is the case of Bare-1 transposon of barley, for which they were developed).

"Single-site" strategies

Single-site strategies are based on rapid detection through PCR of two alternative states, "empty" or "full", of a given site. They were first used in animal systems and more recently in plants, with small retrotransposons such as SINEs, the "empty" or "full" state being manifested in a difference in the size of the PCR product generated from two primers flanking the insertion site. Implementation of single-site strategies requires prior knowledge of the genomic regions flanking the polymorphic insertion sites. Some of the data can be acquired by simple sequencing of polymorphic SSAP fragments extracted from a gel, since they contain flanking sequences

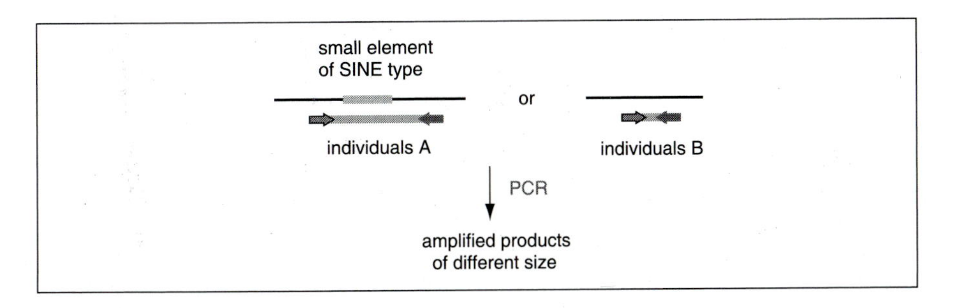

of one side of the element. The flanking sequence of the other side can then be obtained by various strategies, such as for example an SSAP anchored in the known flanking region, and realized on a genotype in which the site is "empty".

Another strategy makes use of a simple PCR between an external flanking primer and an internal primer of the retrotransposon (external/internal PCR). In theory, this strategy does not require knowledge of the size of amplified fragments and is based only on the fact that an amplification product is present ("full" site) or not ("empty" site). It can be implemented by any simple technique detecting the presence of DNA (fluorescent primers, dot blots with probe of insertion site) and thus be automated for the analysis of a large number of samples. An improvement in the strategy was brought about by the development of retrotransposon-based insertion polymorphism (RBIP). RBIP is based on the fact that retrotransposons with LTR are several kilobases long and thus PCR amplification cannot be carried out between two flanking primers since the site is "full". RBIP combines the external/internal PCR mentioned above with a complementary "external/external" PCR between two flanking primers. If the site is "full", only the external/internal PCR will give a product of amplification, and if the site is "empty", only the pair of external primers will give a product of amplification. Each PCR is thus a control on the other, which makes it possible to detect the situations in which the insertion is not fixed in a population and especially to avoid false positives or false negatives.

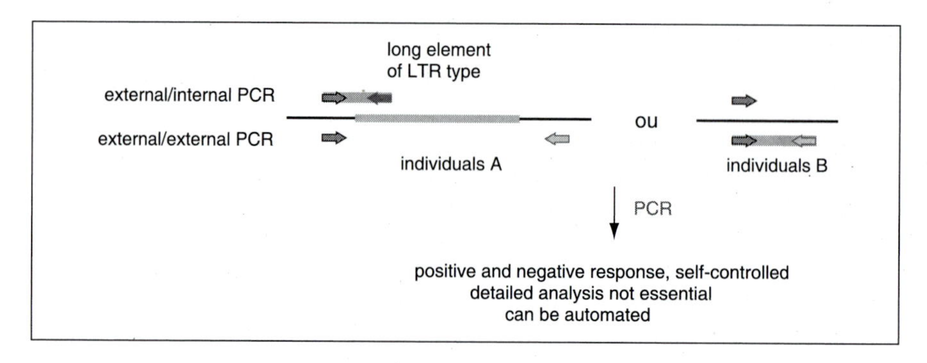

Once these prerequisites are met and the collections of primer patterns obtained, RBIP is a quick and effective tool that can be used for the organization of collections of genetic resources or for protection of varieties or quality control.

Retromarkers are applicable to all species

Retromarkers were first developed with elements that were already known. Nevertheless, they can be considered for any genome, at least for those that involve retrotransposons with LTR. The latter carry extremely conserved proteic motifs in their coding sequences. Degraded primers corresponding to these motifs can theoretically be used to "recover" sequences of retrotransposons from any genome.

Profile 54

SSCP: SINGLE STRAND CONFORMATION POLYMORPHISM

The single strand conformation polymorphism (SSCP) technique is based on the electrophoretic behaviour of a single-stranded DNA fragment in a non-denaturing acrylamide gel. In fact, a molecule of single-stranded nucleic acid may form secondary structures due to pairing of bases within the molecules. These secondary structures depend on the sequence proper to the DNA strand and give a particular conformation to each type of single strand molecule. Thus, two sequences of DNA that are very similar can be differentiated on the basis of the conformation of their single-stranded form. Two alleles of a single gene will be distinguished. This makes it possible, for example, to detect a mutant allele (possibly responsible for a genetic disorder) in an individual. This simple technique has two major disadvantages: the electrophoretic behaviour of single-stranded molecules is unpredictable because it is closely dependent on the temperature and conditions of electrophoresis and, for large fragments (> 200 bp), all the mutations do not seem to be detectable (the method can be used to detect essentially sequence variations of micro-insertion or micro-deletion type).

In practice, the double-stranded product of PCR amplification corresponding to the target sequence is denatured by heating at 94°C, then rapidly cooled in ice: the single-stranded molecules do not have time to anneal with each other but form a stable secondary structure by reassociations at the level of zones of complementary sequences. The reassociated fragments of amplification are then separated by electrophoresis. In a homozygous individual, two fragments are generally observed, because two molecules of complementary single-stranded DNA have slightly different secondary conformations. In a heterozygous individual, four fragments are generally observed. In some cases, extra

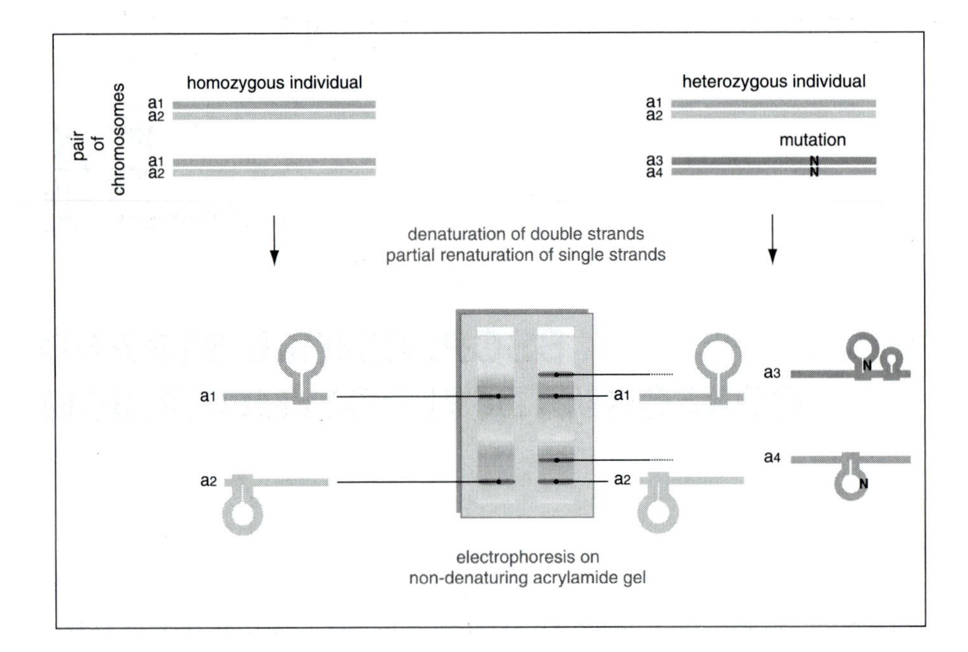

fragments may appear because of the existence of several semi-stable conformations for a single strand.

Profile 55

DGGE: DENATURING GEL GRADIENT ELECTROPHORESIS

Denaturing gel gradient electrophoresis (DGGE) can be used to detect polymorphisms or very fine mutations of DNA, when even a single base pair is substituted in a DNA fragment. It can be applied to fragments of several hundreds of base pairs and does not require prior knowledge of the polymorphic site. This method is also based on the properties of renaturation of different strands of DNA. However, unlike with SSCP (see Profile 54), the denaturation of the DNA occurs during electrophoresis, not before. For this purpose, the products of amplification are deposited on the gel, which contains either an increasing concentration of a chemical agent denaturing the DNA (urea formamide in the case of DGGE) or a temperature that is increasingly higher during the electrophoresis[1] (temperature gel gradient electrophoresis or TGGE). During the electrophoresis, the DNA migrates first to the double strand state, then encounters conditions that denature the least stable molecule with low melting temperature, forming a partly single-stranded structure called "branched" or "ramified". This partial denaturation leads to a reduction of its electrophoretic mobility (the fragment almost stops migrating), so that its final position in the gel depends exclusively on the melting temperature of the least stable field and of the nucleotide sequence of the latter. It is estimated that around 95% of the sequence variations, even point mutations, of the least stable fusion field can be detected by differences in electrophoretic migration.

Moreover, the heterozygous and homozygous individuals can be distinguished by adding a denaturation/renaturation step after the last PCR cycle: a homoduplex (two perfectly complementary strands of DNA) and a heteroduplex (two partly complementary strands of DNA) will form by reassociation of allelic matrices. The occurrence of mismatches in the

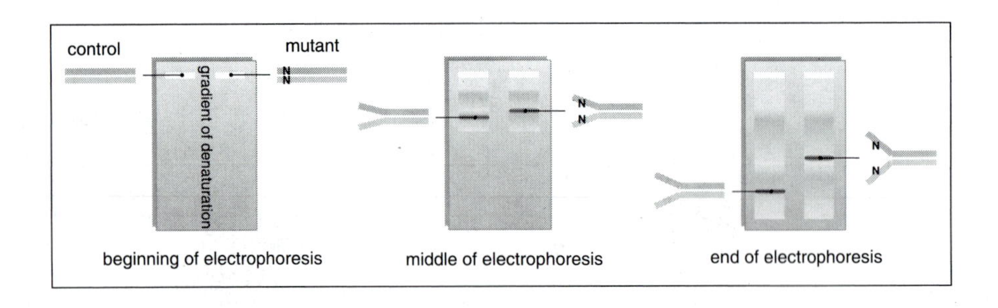

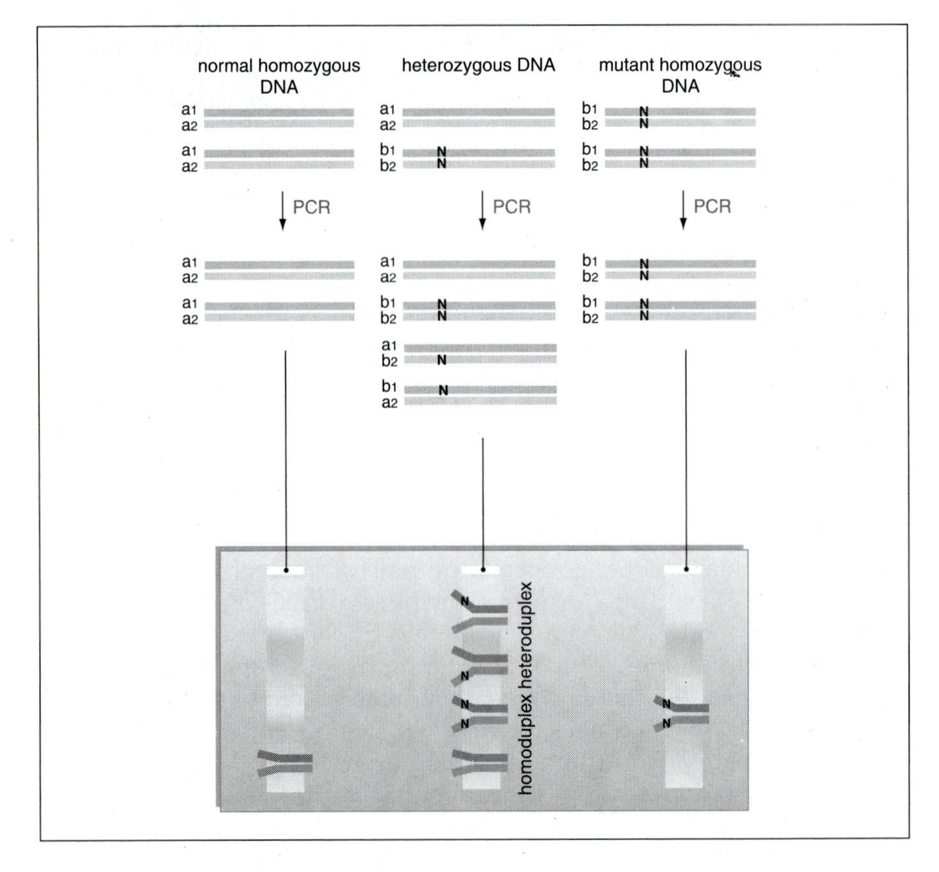

heteroduplex will give rise to highly retained fragments on the gel (rapid denaturation of the hybrid) that hardly migrate.

[1]The gradient of denaturing agents can be converted into a temperature gradient by an empirical formula: T (°C) = (% denaturant + 182.4)/3.2, where 100% denaturant corresponds to a 40% concentration of formamide and 7M of urea.

Profile **56**

SNP: SINGLE NUCLEOTIDE POLYMORPHISM

DNA polymorphism is used as a genetic marker of identification of individuals in many fields (see Profile 47). One such polymorphisms is called single nucleotide polymorphism (SNP). By extension, this name has been given to the markers, SNP markers (pronounced "snips").

Description

There is an occasional isolated mutation that reveals a SNP between two or several DNA sequences. These variations involve a single base that is found randomly around every 100 to 300 bases on a genome. Most SNPs are found in the non-coding part of genomes and, in this case, the polymorphism has no functional implications. However, some of these occasional mutations affect the coding regions (cSNP) and the regulatory regions of genes. They can therefore modify the sequence of amino acids of the corresponding polypeptide or the mode of regulation of the gene expression. The cSNPs are particularly useful in mapping multi-factor diseases through a study of candidate genes involved in such diseases.

Detection

The detection of SNP requires first a PCR amplification of different genotypes. These PCRs can be targeted on known sequences and identified in advance, but they can also be random (AFLP). Second, the different amplification products are sequenced. This technique of SNP requiring the sequencing of PCR products is costly. After sequencing and verification, the different sequences obtained are aligned. The sequence polymorphism sites are then identified, particularly substitution polymorphisms. Thus, the different haplotypes (from different combinations of SNP) can be identified.

Utility

SNPs, like all genetic markers, can be used in different ways. Their extreme precision makes it possible to carry out highly refined analyses in fields such as the search for relationships or variability related to diseases. The efficiency of SNPs is due essentially to their presence in the entire genome, in the introns as well as the exons. Moreover, each individual of a species will have several thousands of SNPs, occasional variations that combine to make each individual a unique "person". The detection of SNP can thus make it possible to discover genes for predisposition to numerous diseases (e.g., diabetes, arterial hypertension, or rheumatoid polyarthritis in humans).

However, the use of SNP involves problems of cost and output of analysis. Some new methods such as mass spectrophotometry or fluorescence make it possible to consider a high-yield genotyping.

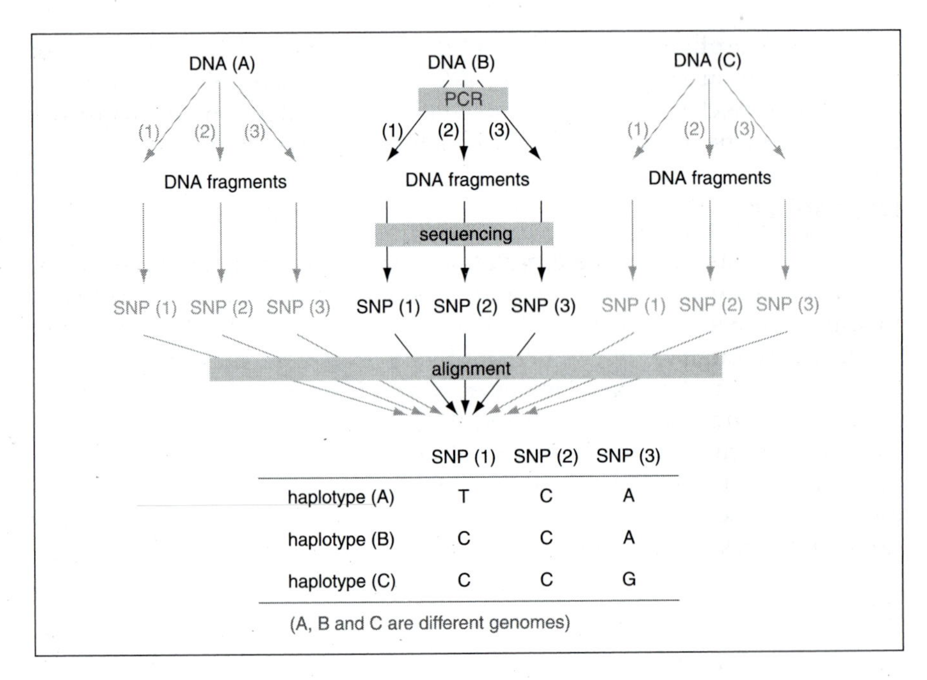

	SNP (1)	SNP (2)	SNP (3)
haplotype (A)	T	C	A
haplotype (B)	C	C	A
haplotype (C)	C	C	G

Profile **57**

SIMPLE SEQUENCE REPEATS (SSR): MICROSATELLITES

Microsatellites, also called simple sequence repeats (SSR), are sequences present in the genomes and are used very often as molecular markers because of their simplicity and wide distribution in genomes of eukaryotic species. They are made up of tandem repeats of a simple DNA motif (a few nucleotides). The motifs may be mono-, di-, tri-, or tetranucleotides, for example CA repeated *n* times for a dinucleotide microsatellite. The peculiarity of microsatellites arises from their use in the search for polymorphism between individuals. This polymorphism does not lie in the sequence itself (CA dinucleotides are present in all genomes) but in the number of repetitions of these simple motifs. For example, at a given locus of a genome, an individual A will have 15 repeats of motif CA, while individual B will have 18 repeats of the same motif.

In order to determine the number of repeats at a given locus, the motif must be specifically amplified by PCR (from genomic DNA of individuals A and B). It is thus necessary to know the flanking sequences of this repetition

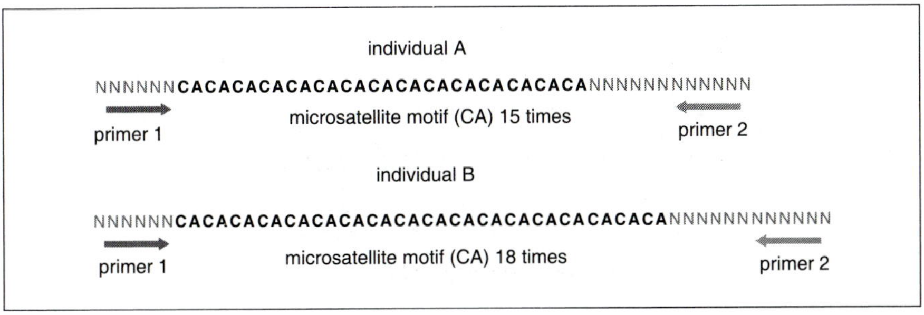

individual A

NNNNNN**CACACACACACACACACACACACACACA**NNNNNNNNNNNN

primer 1 → microsatellite motif (CA) 15 times ← primer 2

individual B

NNNNNN**CACACACACACACACACACACACACACACACACA**NNNNNNNNNNNN

primer 1 → microsatellite motif (CA) 18 times ← primer 2

in order to design primers surrounding the microsatellite motif. Moreover, the fragments of amplification generated differ in size only by a few bases, so a high resolution gel electrophoresis separation is required. The microsatellite locus is thus defined by a pair of PCR primers that amplify a single region on the genome containing a repeat motif.

Use

The use of microsatellites as molecular markers is based on three essential steps: isolation and characterization of microsatellites, designing of specific primers, and search for size polymorphism between individuals.

Isolation of microsatellites is conventionally done in the following steps: (1) creation of a library of short genomic DNA fragments; (2) screening of the library by hybridization; and (3) sequencing of positive clones.

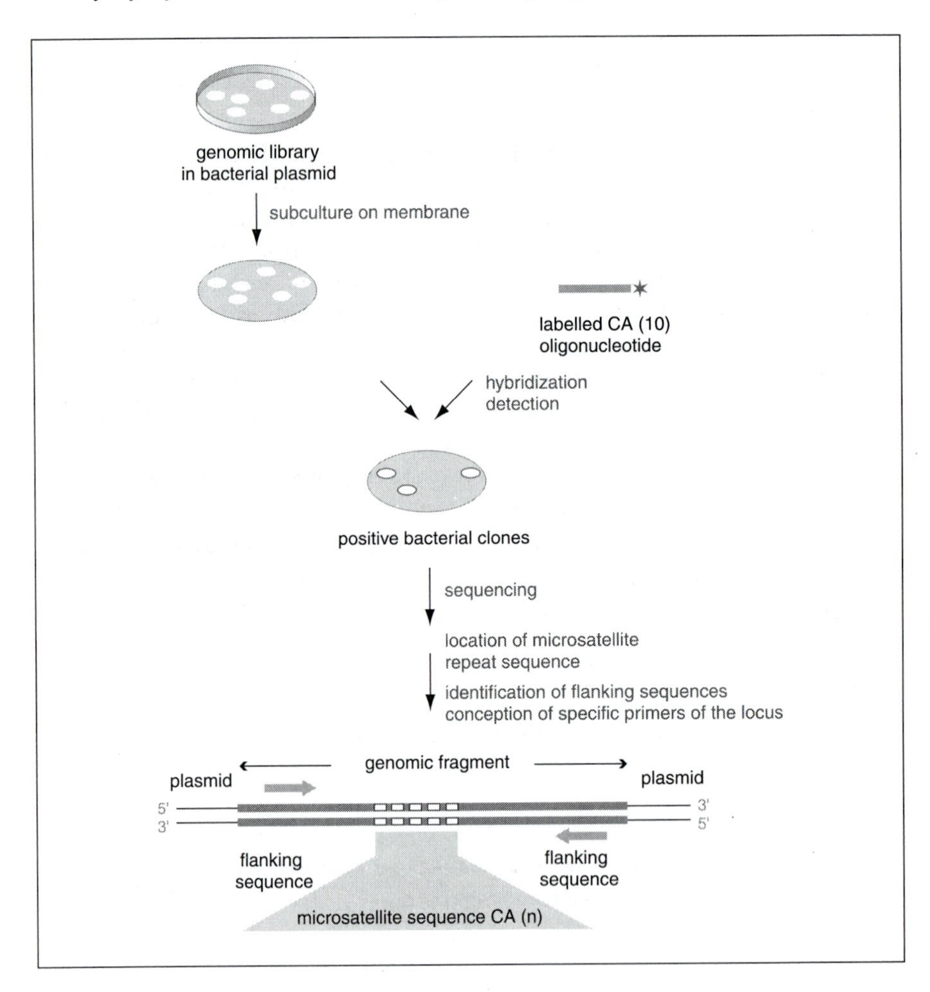

There are two variants:

- Creation of a library enriched in microsatellite motifs, making it possible to skip the steps of screening the library. Several procedures of enrichment are available, but the principle remains the same. The

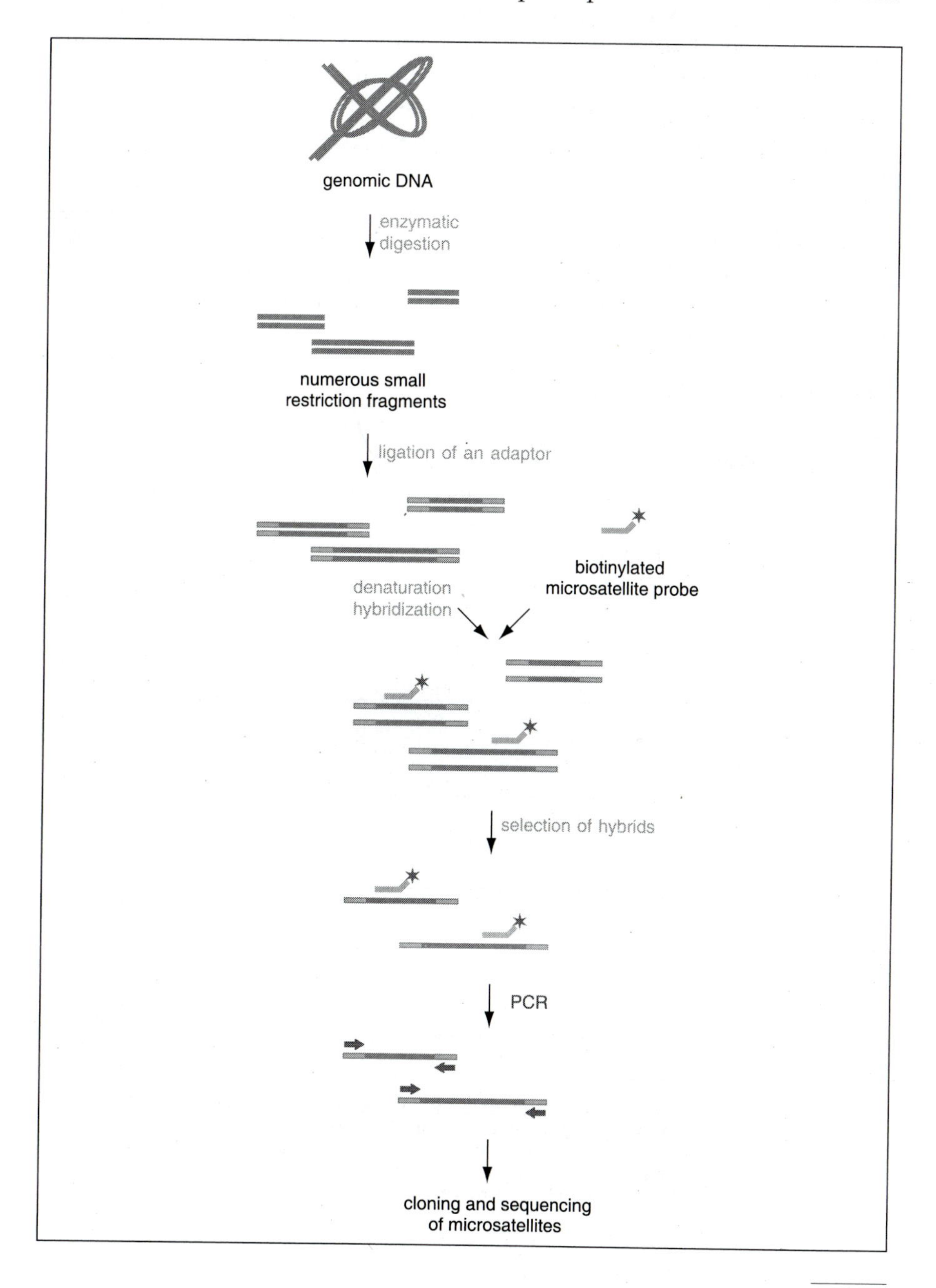

genomic DNA is cleaved then ligated at an adaptor. The fragments are then denatured and hybridized with an oligonucleotide probe of sequence complementary to the microsatellite motif desired. This probe is either biotinylated or fixed on a support (column). Thus, the genomic DNA fragments hybridizing on microsatellite probes are recovered. These purified complexes are then dehybridized and amplified by PCR using complementary primers of the adaptor linked to the ends of genomic DNA fragments.

- Search for microsatellites in genomic data bases or EST.

Designing primers

Primers can be designed in two ways: (1) directly for the cloned sequence obtained or (2) developed on similar species. Amplification assays must then be carried out to verify some characteristics of the microsatellite (presence of null alleles, level of polymorphism).

Detection of polymorphism

After PCR amplification from genomic DNA of different individuals, the products of amplification are separated using gels with resolution adapted to the size differences that need to be revealed. Tri- and tetranucleotide motifs can be separated on high resolution agarose gel. Dinucleotide motifs, which are the most abundant, are generally separated on denaturing

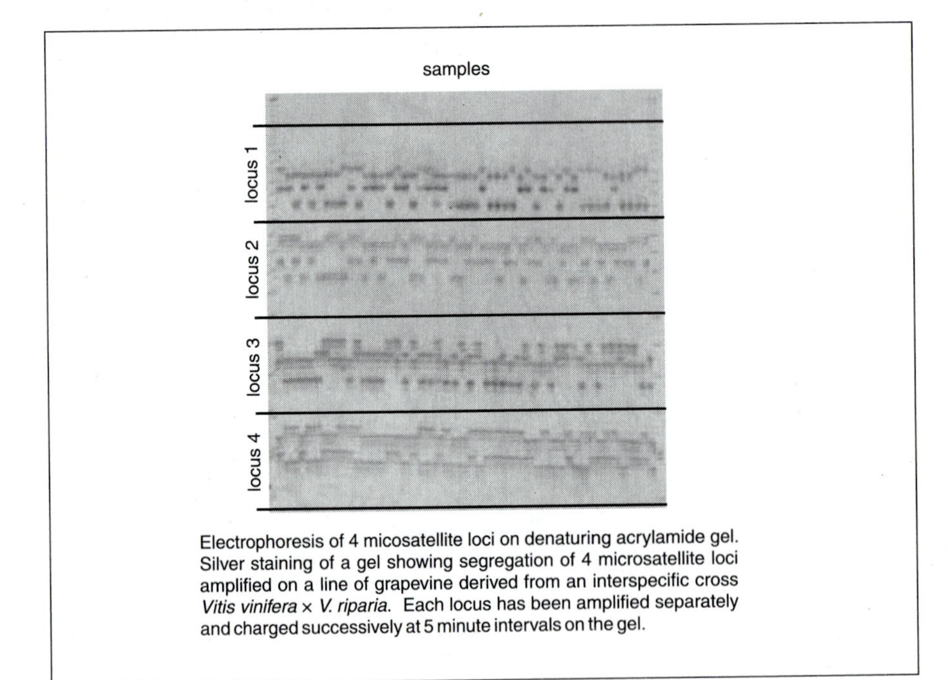

Electrophoresis of 4 micosatellite loci on denaturing acrylamide gel. Silver staining of a gel showing segregation of 4 microsatellite loci amplified on a line of grapevine derived from an interspecific cross *Vitis vinifera* × *V. riparia*. Each locus has been amplified separately and charged successively at 5 minute intervals on the gel.

acrylamide gel. The PCR primers are generally chosen so as to amplify a fragment of size 80 to 200 bp, which allows a clear separation of fragments. In case of automatic sequencer analysis, fragments of size up to 400 bp can be selected. The number of PCR reactions can be reduced by combining several loci in a single tube (multiplexing) but the products of amplification must be of clearly different sizes. Multiplexing can also be carried out by labelling products of similar sizes with different fluorochromes. Since multiplex amplifications are sometimes difficult to carry out, the products can be mixed after PCR.

LIST OF CONTRIBUTORS

Barloy-Hubler Frédérique, Génétique et Développement, UMR 6061, Faculté de Médecine, 2 avenue du Professeur Léon Bernard, 35043 Rennes, France, frederique.hubler@univ-rennes1.fr

Béclin Christophe, IMVT, IBDM, Campus de Luminy, Case 907, 13288 Marseille cedex 09, France, beclin@ibdm.univ-mrs.fr

Cérutti Martine, CNRS, Station de Pathologie Comparée, 30380 Saint-Christol-lès-Alès, France, cerutti@ales.inra.fr

Chalot Michel, Univ. H. Poincaré, Nancy I, Faculté des Sciences et Techniques, IFR 110 Génomique, Ecophysiologie et Ecologie Fonctionnelles, UMR 1136 INRA/Université Nancy I, Interaction Arbres Microorganismes, BP 54506, 54506 Vandoeuvre cedex, France, Michel.Chalot@scbiol.uhp-nancy.fr

Damier Laurence, Unité des Agents Antibactériens, Institut Pasteur, 25 rue du Docteur Roux, 75015 Paris, France, ldamier@Pasteur.fr

Decroocq Stéphane, INRA, UREFV, 71 avenue Edouard-Bourleaux, B.P. 81, 33883 Villenave d'Ornon cedex, France, sdecrooc@Bordeaux.inra.fr

Devauchelle Gérard, CNRS, Station de Pathologie Comparée, 30380 Saint-Christol-lès-Alès, France, devauche@ales.inra.fr

Duplessis Sébastion, UMR 1136 INRA/UHP-Nancy 1 IaM, "Interactions Arbres/Micro-organismes", Centre INRA de Nancy, 54280 Champenoux, France, duplessi@nancy.inra.fr

Frigerio Jean-Marc, INRA, UMR BIOGECO, Site de Recherches Forêt Bois de Pierroton, 69 route d'Arcachon, 33612 Cestas Cedex, France, Frigerio@pierroton.inra.fr

Garcia Virginie, UMR Physiologie et Biotechnologie Végétales, IBVI, INRA Bordeaux, 71 avenue Edouard Bourleaux, 33883 Villenave d'Ornon, France, virginie.garcia@bordeaux.inra.fr

Giblot Danièle, INRA Rennes, UMR INRA/AGROCAMPUS BiO3P, BP 35327, 35653 Le Rheu cedex, France, Daniel.giblot@rennes.inra.fr

Grandbastien Marie-Angèle, INRA, Biologie Cellulaire, Route de Saint-Cyr, 78026 Versailles cedex, France, marie-angele.grandbastien@versailles.inra.fr

Grenier Eric, INRA Rennes, UMR INRA/AGROCAMPUS BiO3p, BP 35327, 35653 Le Rheu cedex, France, eric.grenier@rennes.inra.fr

Hilbert Jean-Louis, UPRES-EA 3569, Physiologie de la Différenciation Végétale, ERT 1016, IFR 118, USTL, 59655 Villeneuve d'Ascq, France, jean-louis.hilbert@unvi-lille1.fr

Houdebine Jean-Marie, UMR INRA/ENV Alfort/CNRS Biologie du Développement et Reproduction, INRA Domaine de Vilvert, 78352 Jouy-en-Josas cedex, France, Louis.Houdebine@jouy.inra.fr

Hugot Karine, UMR INRA/CEA Radiologie et Etude du Génome, INRA Domaine de Vilvert, 78352 Jouy-en-Josas cedex, France, karine.hugot@jouy.inra.fr

Jacquot Jean-Pierre, Univ. H. Poincaré, Nancy I, Faculté des Sciences et Techniques, UMR 1136 INRA/Université Nancy I, Interaction Arbres Microorganismes, BP 54506, 54506 Vandoeuvre cedex, France, jean-pierre.jacquot@scbiol.uhp-nancy.fr

Leblond Pierre, UMR INRA 1128 Génétique et Microbilogie, Univ. H. Poincaré, Nancy I, Faculté des Sciences et Techniques, BP 54506, 54506 Vandoeuvre cedex, France, pierre.leblond@scbiol.uhp-nancy.fr

Leplé Jean-Charles, INRA Orléans, Amélioration Génétique et Physiologie Forestière, BP20619 Ardon, 45166 Olivet cedex, France, leple@Orleans.inra.fr

Massonneau Agnès, RNA works, Cap Alpha, Avenue de l'Europe, 34940 Montpellier cedex 09, France, a.massonneau@rnaworks.net

Méreau Agnès, UMR 6061 Génétique et Développement, Faculté de Médecine, 2 avenue du Professeur Léon Bernard, 35043 Rennes, France, agnes.mereau@univ-rennes1.fr

Françoise Monéger, Reproduction et Développement des Plantes, ENS Lyon, IFR 128, 46 Allée d'Italie, 69364 LYON Cedex 07, France, Francoise.Moneger@ens-lyon.fr

Moussard Christian, Service de Biochimie Médicale, Faculté de Médecine, Place Saint-Jacques, 25000 Besançon, France, christian.moussard@ufc-chu.univ-fcomte.fr

Pilate Gilles, INRA Orléans, Amélioration Génétique et Physiologie Forestière, BP20619 Ardon, 45166 Olivet cedex, France, pilate@Orleans.inra.fr

Plomion Christophe, INRA, UMR BIOGECO, Equipe de Génétique, Site de Recherches Forêt Bois de Pierroton, 69 route d'Arcachon, 33612 CESTAS Cedex, France, christophe.plomion@pierroton.inra.fr

Samson Manuella, 5 rue de I'Yser, 92210 Saint-Cloud, France, manuella.samson@edfgdf.fr

Sourdille Jean-Pierre, UMR INRA-Université Clermont II, Amélioration et Santé des Plantes, site de Crouël, 234 avenue du Brézet, 63100 Clermont-Ferrand cedex 2, France, pierre.sourdille@clermont.inra.fr

Tagu Denis, INRA Rennes, UMR INRA/AGROCAMPUS BiO3P, BP 35327, 35653 Le Rheu cedex, France, denis.tagu@rennes.inra.fr